MAP PROJECTIONS

MAP PROJECTIONS

Hugh S. Roblin, B.A.

Lecturer in Geography at Cardiff College of Education
Formerly Head of the Geography and Geology Department
at Howardian High School for Boys, Cardiff

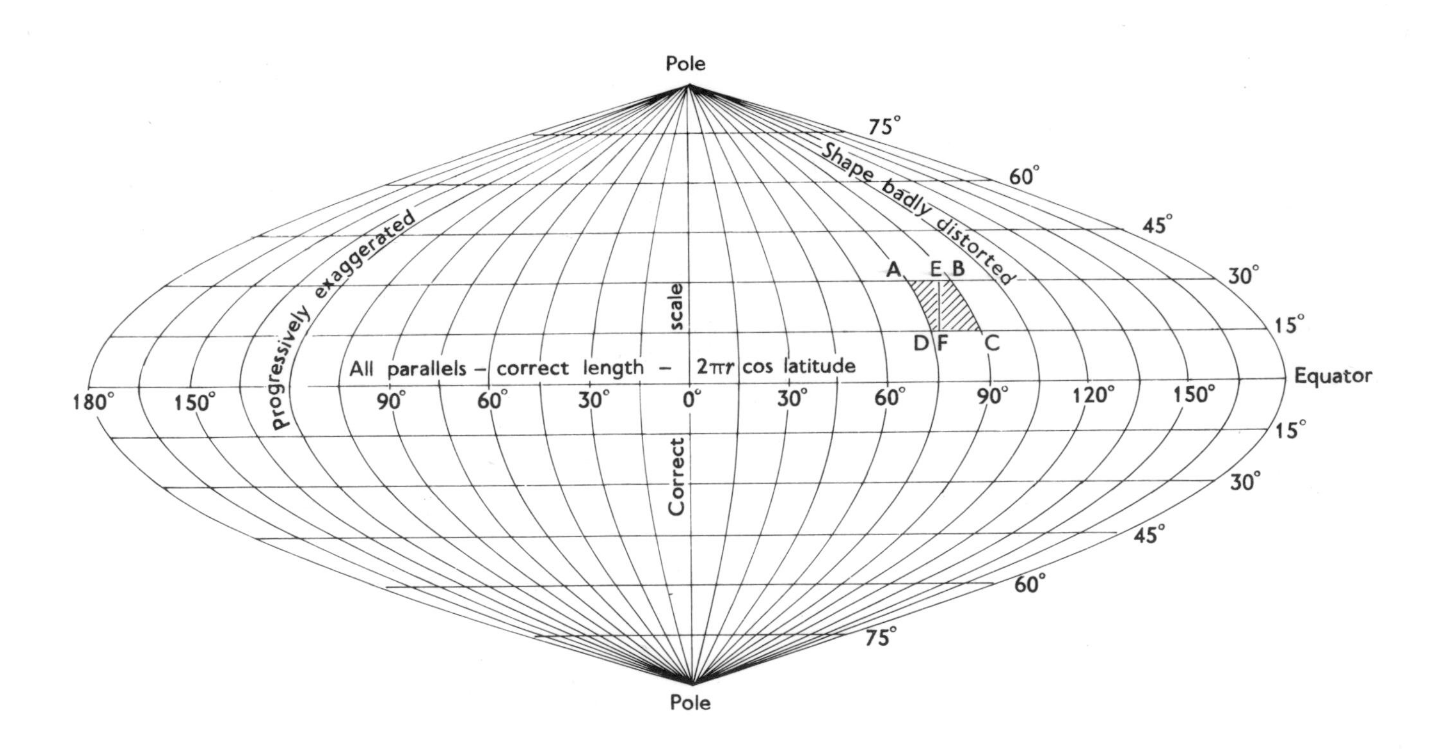

EDWARD ARNOLD (Publishers) LTD.

First published 1969

SBN : 7131 1528 9

Printed in Great Britain by
Fletcher & Son Ltd, Norwich

PREFACE

This book aims at providing a simple, straightforward explanation of the map projections in common use in modern atlases. Although it is primarily intended for 'A' Level pupils and students in Colleges of Education, it will form a suitable background to the more advanced study often undertaken in the first year of a university course.

Geography students are seldom mathematicians and some are often unnecessarily discouraged by the appearance of even the simplest mathematics in textbooks. For this reason, mathematical explanations are avoided, as far as possible, in the description and discussion of each projection. They are given, where necessary, under the title 'Calculation' at the end of the relevant chapter, so that those students who really want to,can gain some insight into the simple mathematics behind most of the commoner map projections.

An essential feature of the book is the table in Chapter 2 which shows at a glance under convenient headings the major characteristics, properties and uses of each projection dealt with.

It cannot be urged too strongly that students should, when reading this or any other book on map projections, consult an atlas or atlases containing a variety of projections to test for themselves, on actual examples, the validity of any statements concerning each.

H.S.R.

CONTENTS

MAPS AND DIAGRAMS

PART I

1 The Problem of Map Projection

The globe is a true representation of the earth and, provided it is large enough, all parts of the earth's surface can be represented on it in their true shape, relative size and position; but great problems arise when an attempt is made to transfer details of land-masses and oceans from the spherical surface of the globe to the plane surface of a sheet of paper. It is impossible to do it with complete accuracy of shape, relative size and position at the same time, so that the study of map projection resolves itself into a study in compromise.

Because the position of any point on the earth's surface can accurately be defined in terms of its latitude and longitude, the system of meridians and parallels is the framework, or skeleton of the land-masses. The study of map projection is the study of the various ways which may be devised for transferring these meridians and parallels from the spherical surface of the globe to the plane surface of a sheet of paper. Clearly, it is impossible to make the 'skin' of a globe lie flat without stretching it in places. It is possible to devise a framework of intersecting meridians and parallels so that land-masses plotted on them are truly represented in area, but only by the sacrifice of their shape. Conversely, true shape can be preserved only at the expense of area and, even then, one cannot on a plane surface faithfully represent the shape of a *large* part of the earth's surface. A map can preserve the shape of *small* areas, strictly speaking of *infinitely small* areas, and it is then said to be 'orthomorphic' (Greek 'orthos'—straight or correct: 'morphos'—shape).

The particular system or projection adopted for representing parallels and meridians will depend on the purpose for which the final map is required. Some projections show equatorial areas well, if not with complete accuracy of shape or area, while polar areas are badly distorted. Others preserve area and not shape and so are best used for showing distributions, for example, of races or rainfall or natural vegetation; areas under coniferous forest may thus be compared visually with that under equatorial forest or hot desert or steppe.

The Problem of Scale

The earth is not an exact sphere; it has an *equatorial* diameter of 7,927 miles (radius 3,963·5 miles) but a *polar* diameter of 7,920 miles (radius 3,950 miles). However, for the purposes of map projection, it can be regarded as being spherical because, although the equatorial circumference (24,891 miles) is 71 miles greater than the polar circumference (24,820 miles), this difference becomes insignificant when applied to a globe or reduced earth; it therefore becomes unnecessary, if not impossible to attempt to differentiate between one dimension and the other.

What is important in map projection is to decide on the *radius of the globe* which is going to represent the *radius of the earth*—that is, to decide on a scale for the globe from which the map projection is to be developed. When the globe has been constructed on this scale, every part of the globe will have the *same* reduced scale relationship to the corresponding part of the earth. The equator, all other parallels, and all the meridians will bear the same scale relationship to those on the earth and each will be of correct length in relation to all the others. This is not so on a map projection. As has been said, it is impossible to make the 'skin' of a globe lie flat without stretching it in places or causing it to shrink in others relative to their counterparts on the globe. Some parallels or meridians may be shorter than those on the globe, i.e. they are *reduced in scale*; others may be longer than those on the globe, i.e. they may be exaggerated in length or, put another way, they are *increased in scale*. Whereas the globe as a whole, therefore, has a reduced scale relationship to the earth it represents, only certain meridians or certain parallels on a map projection representing the meridians and parallels of the globe preserve this scale relationship. For example, when a scale is placed beneath a map, it refers to lengths or distances on certain parts of it *only*.

On the well-known Mercator projection, the scale which is printed on it is the *equatorial* scale, because of all the meridians and parallels on it, the equator is the only one which has the same length as its counterpart on the globe which is represented. To take another example, on the familiar elliptical projection known as Mollweide's, the equator is too short and only *one* of the other parallels which could be put on it would be correct; they are all either longer or shorter than their counterparts on the globe; similarly for the meridians.

It often becomes meaningless, therefore, to quote a scale in terms of an inch on the map to so many miles on the earth. Then the only meaningful thing to do is to quote the scale relationship between the radius of the globe which the projection represents and the actual radius of the earth—in terms of a representative fraction (see below). When this is known and the scale characteristics of a projection are known, it is then often a matter of simple arithmetic to find out the number of times by which a parallel or a meridian is either too short or too long compared with their counterparts on the globe. The exaggeration (or diminution) of any meridian (or parallel) on the projection can be calculated as follows:

$$\frac{\text{Length on projection}}{\text{Length on globe}}$$

If multiplied by 100, this becomes a percentage of the correct length.

The Representative Fraction

The equatorial radius of the earth—3,963·5 miles, expressed in inches is 251,117,360. The scale of a globe, therefore, of radius 1 inch, expressed as a representative fraction is 1 : 251,117,360 or approximately 1 : 250 million.

The approximate representative fraction (R.F.) of a globe of 10 inches radius is 1 : 25 million, of 100 inches radius 1 : 2·5 million and of 250 inches radius 1 : 1 million.

Shape *v* Area

The Maintenance of True Shape

One of the major characteristics of meridians and parallels on the globe is that they intersect at right angles. If this right-angled intersection is not maintained on a projection, under no circumstances can the shape of a land-mass be maintained, even though all other dimensions remain true to scale. This can be illustrated as follows:

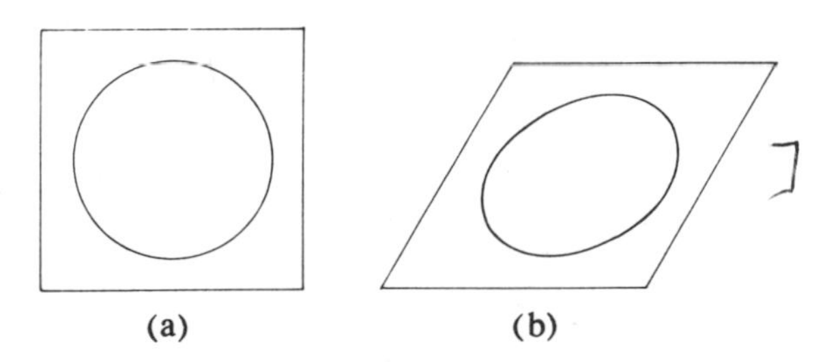

Fig. 1 To illustrate the necessity for retaining the right-angled intersection.

Fig. 1(a) represents a picture frame which can be swivelled at its corners, the picture itself being a circle printed on a rubber sheet secured to the frame. Fig. 1(*b*) shows what happens to the circle when the frame is pulled out of square, a situation analogous to the effect on the shape of land-masses by the failure to preserve the right-angled intersection of meridian and parallel on the projection.

Fig. 2(*a*) is the actual shape on the globe of a very small area of land, 1 unit × 1 unit. If, on the projection, the scale along the meridian is doubled relative to that on the globe while the parallel scale remains correct (Fig. 2(*b*)) the square will be stretched, north–south, to become a rectangle, 2 units × 1 unit. The shape of the original square has been destroyed and its area doubled. Similarly, if the parallel scale is doubled while the meridian scale remains correct, the square shape has again been destroyed and the land-mass is represented by a rectangle, 1 unit × 2 units: the area has again been doubled (Fig. 2(*c*)). But if, as in Fig. 2(*d*) meridian and parallel scales are both doubled, while the right-angled intersection is maintained, the original square will be represented by another square. Its shape has been preserved even though its area has been quadrupled.

The Maintenance of Area

Consider the same square land-mass of side 1 unit. If the meridian scale is halved while the parallel scale is doubled, the square land-mass will turn out to be a rectangle 2 units × $\frac{1}{2}$ unit (Fig. 3). This embodies the principle of the maintenance of area by compensatory scales, whereby any exaggeration of the parallel scale is balanced at any point by an equivalent diminution of the meridian scale, always provided that the right-angled intersection of meridian and parallel is maintained; this is the principle underlying the Lambert's Cylindrical Equal-area (or Cylindrical Orthographic) projection. (See Part II, Chapter 4.)

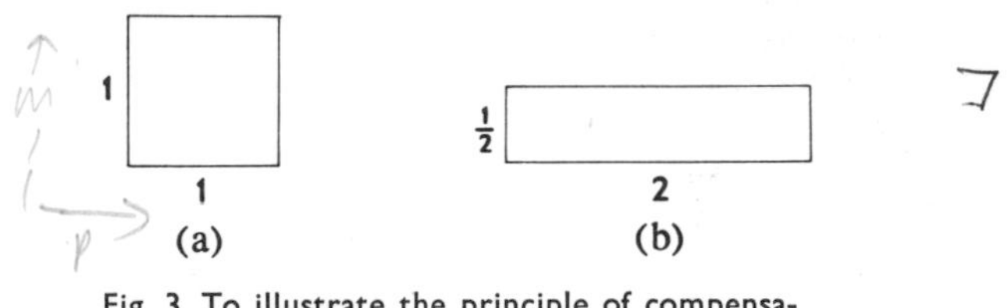

Fig. 3 To illustrate the principle of compensatory scales.

Area can also be maintained where meridians and parallels do *not* intersect at right angles. The square land-mass of side 1 unit is the same in area as a rhombus in which a pair of opposite sides are both 1 unit in length, provided that the perpendicular distance between them is also 1 unit (Fig. 4).

A more common geometrical figure formed by a pair of meridians and a pair of parallels on a map projection is the trapezium. The area of a trapezium is the average length of the parallel sides × the perpendicular distance between them, i.e. $\frac{AB+DC}{2} \times PQ$ (Fig. 5.)

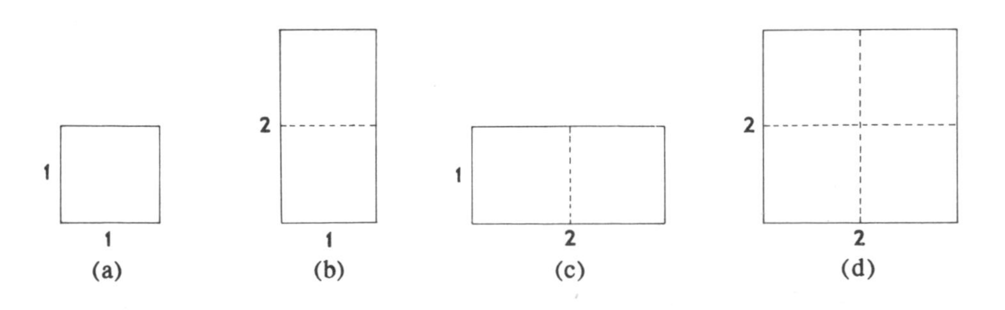

Fig. 2 To illustrate the necessity for maintaining the equality of meridian and parallel scales in order to retain shape.

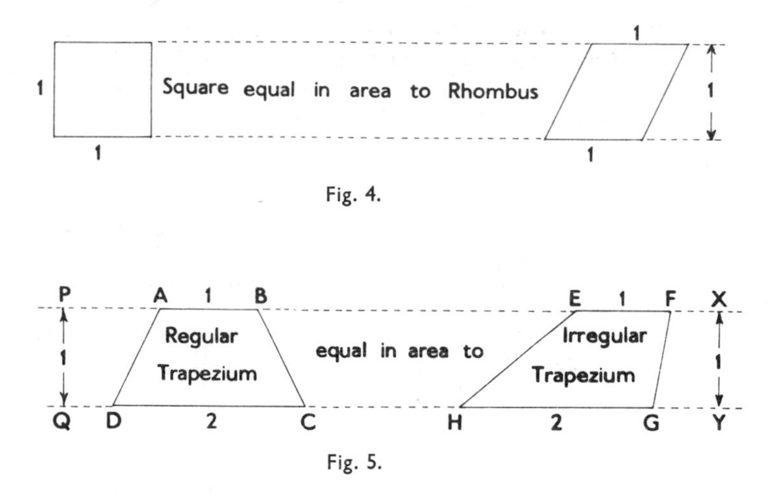

Fig. 4.

Fig. 5.

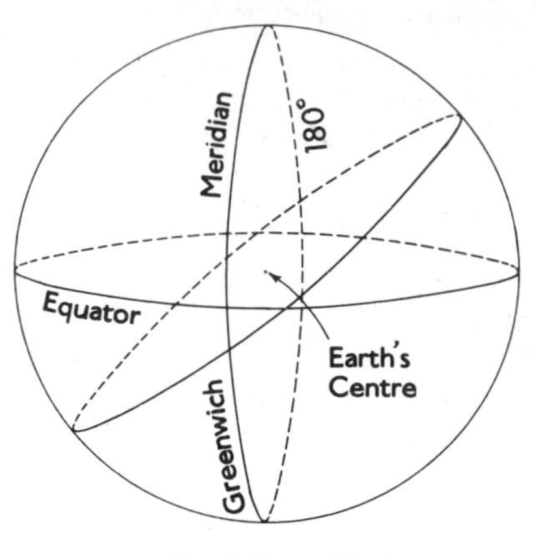

Fig. 6 Great Circles.

The areas of the two trapezia shown are equal because the *significant* dimensions are the same in both: $AB=EF$, $DC=HG$, $PQ=XY$. This simple geometrical fact is important in the maintenance of area in map projection because any pair of intersecting meridians and parallels on the globe forms a regular trapezium.[1] It is the basis of two important equal-area projections to be studied later—Bonne's (Part III, Chapter 9) and Sanson–Flamsteed (Part IV, Chapter 10).

[1] Examine the globe to establish this for yourself. Note that each trapezium on the globe has four right angles. The fact that each trapezium on the globe is made up of curved lines does *not* invalidate the principle.

Great and Small Circles

A great circle is a line drawn on the globe so that it cuts the globe in half. Therefore, the plane of the great circle passes through the centre of the earth and the circle itself is the maximum circumference of the globe. The equator is, thus, a great circle and so is any pair of meridians 180° apart; but a great circle can be drawn obliquely to the meridians and parallels through any two chosen points, and the circle so drawn will be the shortest distance between the two points. (Fig. 6.)

Consider, for example, any two points on the same parallel of latitude; the shortest distance between them is *not* along their common parallel of latitude but along the *only* great circle which can be drawn through both. This is easily demonstrated on the globe. Mark two points on the same parallel of latitude; a piece of string held tightly between them will assume a great circle course which will obviously be a shorter distance than that along the parallel of latitude (or *small* circle) between the two points. The greater the longitudinal distance between the two points, the more obviously shorter is the great circle course. It will be further noticed that the great circle course lies polewards of the small circle; this can be checked by selecting two similar points in the opposite hemisphere. (See Fig. 7.)

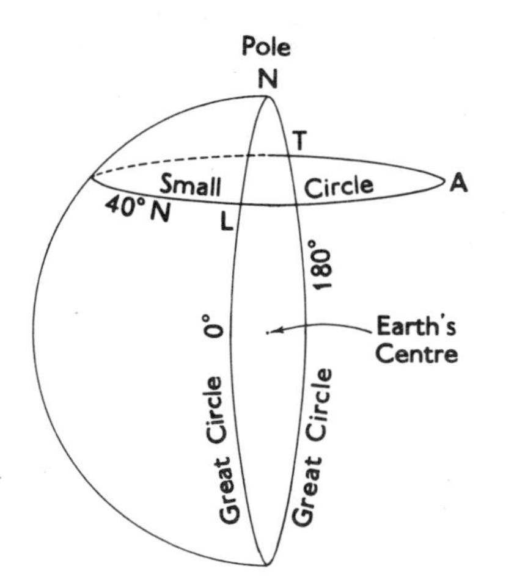

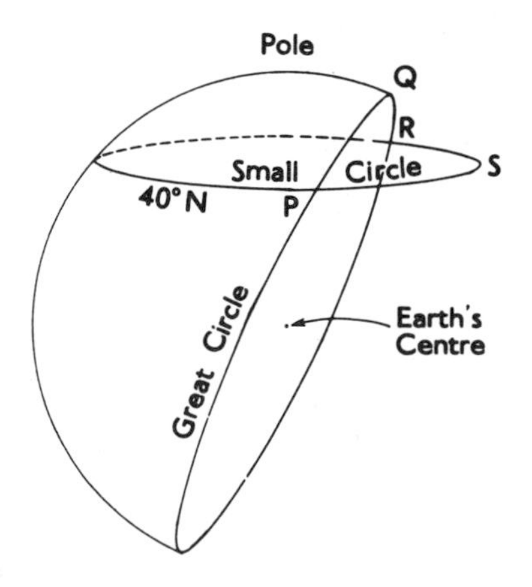

Fig. 7 Great and Small Circles.

(*a*) Where the great circle is a double meridian.
LNT is the great circle distance.
LAT is the small circle distance.

(*b*) Where the great circle is oblique to the meridians and parallels.
PQR is the great circle distance
PSR is the small circle distance.

Measurements on the Globe

It is essential to be able to make some simple calculations of the lengths of parallels and meridians on the globe. (Fig. 8.)

(*a*) Length of the equator on a globe of radius r $= 2\pi r$
(*b*) Length of a great circle $= 2\pi r$
(*c*) Length of a meridian $= \pi r$
(*d*) Length of a parallel of latitude $= 2\pi r$ cosine latitude.

(*For proof, see Fig.* 9.)

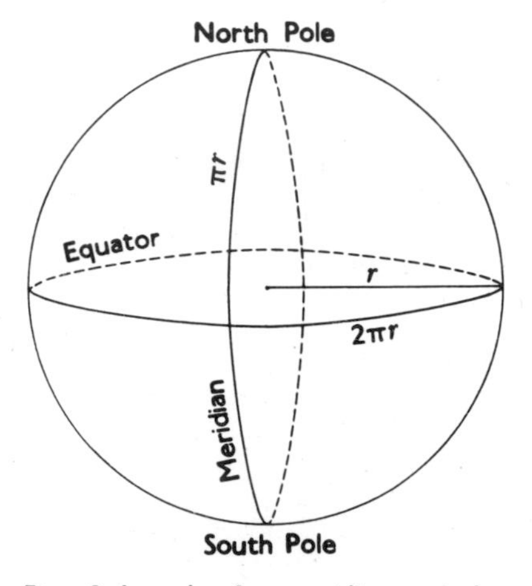

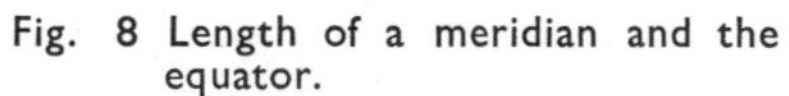
Fig. 8 Length of a meridian and the equator.

Fig. 9 represents a globe whose radius is r. C is its centre and LT the radius of latitude 60°N. To find the length of 60°N. in terms of r:

$\angle TCQ$ (angle of latitude) $= \angle LTC$ (alternate angles)

In $\triangle LTC$ $\frac{LT}{CT} = \text{cosine} \angle LTC$

Therefore, $LT = CT \times \text{cosine} \angle LTC = r \cos$ latitude 60°

Therefore, length of latitude 60° N. $= 2\pi r \cos 60°$
i.e. $2\pi r$ (length of equator) $\times$ cosine of angle of latitude, 60°N.
Similarly for any other parallel of latitude.

Examples

(*a*) Length of parallel 60°N. $= 2\pi r \cos 60 = 2\pi r \times 0{\cdot}5 = \pi r$,
i.e. parallel 60°N. $= \frac{1}{2}$ length of equator.
(*b*) 90°N. (the pole) $= 2\pi r \cos 90° = 2\pi r \times 0 = 0$,
i.e. the pole is a point.
(*c*) Length of 15° of longitude along a selected parallel, e.g. 30°N.:
Length of 30°N. $= 2\pi r \cos 30$.

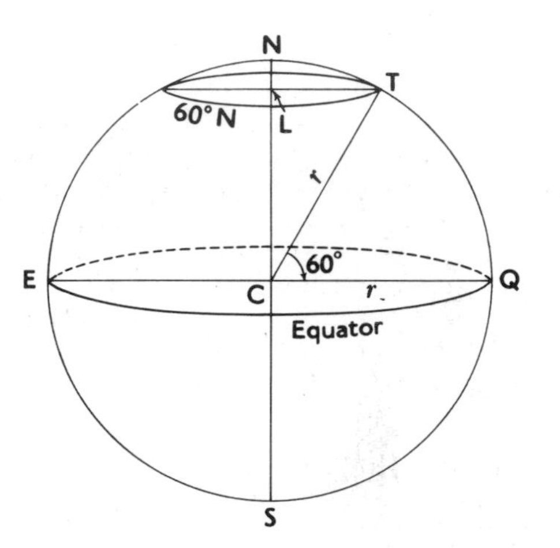

Fig. 9 The length of a parallel of latitude.

Length of 1° of longitude along 30°N. $= \frac{2\pi r \cos 30}{360}$

Length of 15° of longitude along 30° N. $= \frac{2\pi r \cos 30}{360} \times 15$

$= \frac{2\pi r \cos 30}{24} = \frac{\pi r \cos 30}{12}$

2 Summary of Characteristics, Properties and Uses

PROJECTION	SHAPE OF PARALLELS	SHAPE OF MERIDIANS, etc.	INTER-SECTION OF MERIDIANS AND PARALLELS	SCALE ALONG MERIDIANS	SCALE ALONG PARALLELS	REPRE-SENTATION OF SHAPE	REPRE-SENTATION OF AREA	OTHER PROPERTIES	USES
1. SIMPLE CYLINDRICAL	Parallel straight lines	Parallel straight lines	Meridians and parallels cut at right angles	All correct	Equator correct; all others exaggerated as secant of latitude	Not orthomorphic. Tropical areas are of reasonable shape. Land-masses stretched E.–W. in higher latitudes	Not equal-area. Areas progressively exaggerated polewards	Direction not correctly represented	Use best restricted to tropical areas. Rarely used. Equal-area projections more suitable
2. CYLINDRICAL EQUAL-AREA (LAMBERT'S)	Parallel straight lines	Parallel straight lines	Meridians and parallels cut at right angles	Diminishes polewards. At any point, it is as much too small as the parallel scale is too big; i.e. meridian and parallel scales are compensatory	Equator correct; all others exaggerated as secant of latitude	Badly distorted polewards of 45° latitude. Land-masses elongated E.–W., but compressed N.–S.	Equal-area; meridian and parallel scales compensatory	Direction not correctly represented	Distribution maps in tropical areas, but Sanson–Flamsteed's or Mollweide preferred
3. MERCATOR (CYLINDRICAL ORTHOMORPHIC)	Parallel straight lines	Parallel straight lines	Meridians and parallels cut at right angles	Increases progressively polewards in the same ratio as the exaggeration of the parallel scale	Equator correct; all others exaggerated polewards as secant of latitude	Orthomorphic, i.e. correct shape for infinitely small areas. Large land-masses with great extent in latitude are 'top-heavy'	Area is greatly exaggerated polewards, as the square of the secant of the latitude	Any straight line is a line of constant bearing, i.e. a rhumb-line or loxodrome. Great Circles are curved lines convex polewards	Especially suitable for air and sea navigation or any purpose for which representation of direction is required, e.g. wind direction or ocean currents
4. GALL'S PROJECTION	Parallel straight lines	Parallel straight lines	Meridians and parallels cut at right angles	Increases progressively polewards, but polewards of 45° latitude, scale is too	Correct *only* along 45° parallels. Polewards of 45° latitude, very much too	Land-masses in high latitudes are stretched E.–W.; they are also stretched	Areas progressively diminished equatorwards of 45° latitude, but progressively	Direction not correctly represented. Compromise between accuracy of	Gives reasonable general-purpose map of world, except in high latitudes. An equal-area

				wards of 45° latitude, meridian scale is too small	wards of 45° latitude, too small. Equator is 7/10ths of its actual length	smaller extent. Therefore, land-masses appear to be compres-sed. Northern continents 'top-heavy'	wards of 45° latitude		preferred for distributions
5. CONICAL WITH ONE STANDARD PARALLEL (SIMPLE CONICAL)	Concentric circles, the correct dis-tance apart. The pole is represented by an arc of a circle	Straight lines converging on the centre of curvature of the paral-lels	Meridians and parallels cut at right angles	All correct	Correct along standard parallel. All other parallels are too long. Exaggeration is progressively increased pole-wards and equatorwards of the standard parallel	Land-masses are increasingly badly distorted away from the standard parallel. They are badly stretched E.–W.	Exaggeration of area increases rapidly pole-wards and equatorwards of the standard parallel	Direction is not correctly represented	Cannot be used for areas of great extent in latitude. Suitable only for small countries, with not more than 10° extent from N.–S. The conical with *Two* Stan-dard parallels preferred
6. CONICAL PROJECTION WITH TWO STANDARD PARALLELS	Concentric circles, the correct dis-tance apart. The pole is represented by an arc of a circle	Straight lines converging on the centre of curvature of the parallels	Meridians and parallels cut at right angles	All correct	Scale along the two standard parallels is correct. Those between the two standards are too short. Outside the standards, the scale increases progressively	Not orthomor-phic. Although an improvement on the one Standard Parallel Conic, land-masses are elongated E.–W. progres-sively polewards and equator-wards of the standard parallels	Not equal-area. Increasing exaggeration of area polewards and equator-wards of the standard parallels	Direction is not correctly represented	An improvement the one Standard Parallel Conical. Error of parallel scale is more evenly spread, but areas are increasingly exaggerated beyond the standard parallels as the meridians become increas-ingly elongated E.–W. Therefore small countries
7. BONNE'S PROJECTION	Concentric circles, the correct distance apart. The pole is represented by a point 	Composite curves, *not* arcs of circles	Only the central meridian cuts the parallels at right angles. Obliquity between meridian and parallel increases towards mar-gins of map	Increases pro-gressively towards the margins of the map, especially in middle and high latitudes	All correct	Shape of land-masses deter-iorates with increasing distance from the central meridian, especially in middle and high latitudes	Equal-area	Direction is not correctly represented	Suitable for land-masses in one hemisphere, provided that E.W. extent is not too great. General-purpose and distribution maps of contin-ents (or parts), e.g. North America, South America, Australia, Europe, but not Eurasia

PROJECTION	SHAPE OF PARALLELS	SHAPE OF MERIDIANS, etc.	INTER-SECTION OF MERIDIANS AND PARALLELS	SCALE ALONG MERIDIANS	SCALE ALONG PARALLELS	REPRE-SENTATION OF SHAPE	REPRE-SENTATION OF AREA	OTHER PROPERTIES	USES
8. THE POLYCONIC PROJECTION (Infrequently used in atlases and hence not discussed later in the book)	Arcs of circles, but *not* concentric. Each parallel has its own radius	Curved lines but *not* arcs of circles	At right angles *only* along the central meridian. Increasing obliquity away from the central meridian	Correct only along the central meridian. Increasingly exaggerated away from the central meridian	All correct	Shape badly distorted as meridians become progressively elongated away from the central meridian	Not equal-area. Areas increasingly exaggerated away from the central meridian	Direction is not correctly represented	Suitable for relief maps, but only for small areas. Is the basis of the *International Map* on a scale of one in a million. Not often used in atlases
9. SANSON–FLAMSTEED'S SINUSOIDAL PROJECTION	Straight lines of correct length and correct distance apart	All except the central meridian are composite curves	Only the central meridian cuts the parallels at right angles. All others are increasingly oblique to the parallels	Increasingly exaggerated East and West of the central meridian	All correct	Very bad peripheral distortion of shape of land-masses. Extreme eastern and western margins elongated and 'pulled' out of upright. Improved by interrupting	Equal-area	Direction is not correctly represented	Seldom used for the whole globe without interruption. Gives a good equal-area map of continents lying astride the equator with relatively small E.–W. extent, e.g. Africa, South America
10. MOLLWEIDE'S HOMOLOGRAPHIC PROJECTION	Straight lines becoming closer together polewards	Ellipses, *except* central meridian (a straight line) and 90° E.–W. (semicircles)	Only the central meridian cuts the parallels at right angles. All others are increasingly oblique to the parallels towards E.–W. margins	Central meridian is too short. Increases away from central and eventually becomes progressively exaggerated	Equator and other parallels to about 45° latitude are too short. Between about 45° latitude and the pole, parallels are too long	Bad peripheral distortion of shape of land-masses, but better than on Sanson–Flamsteed. Shape within 30° of the central meridian is good	Equal-area	Direction is not correctly represented	Peripheral distortion of shape handicaps use for whole globe; but interruption and recentring improves shape of land-masses. Philip's *University Atlas* uses interrupted case for world distributions
11. POLAR ZENITHAL GNOMONIC	Concentric circles with	Straight lines their	Right angle	Progressively increases away	Progressively exaggerated	Reasonable shape within	Areas of land-masses progres-	(*a*) Direction from the centre	Navigational and general-purpose

	[illegible]	[illegible] distance apart, and radiating from the pole		10% exaggeration at 60° latitude, over 27% at 45° latitude	pole; 15% at 60° latitude, over 40% at 45° latitude	but equatorwards of about 60° latitude, very rapid and great elongation of land-masses from E.–W.	gerated equatorwards. Exaggeration is pronounced equatorwards of 60° latitude	correct. (*b*) Any straight line drawn on the map is a great circle	areas
12. EQUATORIAL ZENITHAL GNOMONIC	The equator, a great circle is, represented by a straight line. Other parallels are composite curves, more markedly curved polewards	Parallel straight lines at right angles to the equator. N.B. *all* great circles, including meridians are represented by straight lines	Equator cuts straight meridians at right angles. Parallels cut meridians increasingly obliquely: (*a*) towards the margins of the map, (*b*) polewards	Increases progressively polewards. Exaggeration along successive meridians is progressively greater E. and W. of the central meridian	Progressive exaggeration polewards	Land-masses elongated progressively N.–S. and E.–W. away from the equator and the central meridian respectively. Shape is reasonably represented within 35° of both the equator and the central meridian	Land-masses exaggerated in area progressively N.–S. and E.–W. away from the equator and the central meridian respectively. Area is reasonably represented within 35° of both the equator and the central meridian	As in 11 above: (*a*) Direction from the centre of the map is correct. (*b*) Any straight line drawn on the map is a great circle	Gives good representation of land-masses near the centre of the projection, provided they do not extend much more than 30° in any direction. Because of correct direction from the centre and representation of a great circle by a straight line, it is suitable for Africa and tropical South America
13. OBLIQUE ZENITHAL GNOMONIC	The equator, being a great circle, is represented by a straight line. Other parallels are composite curves more markedly curved polewards	Being great circles, they are represented by straight lines, converging on the pole which is a point	Only the central meridian cuts the parallels at right angles. Other meridians cut the parallels increasingly obliquely towards the margins of the map, especially within 30° of the central meridian	Scale along the central meridian increases uniformly towards the equator and the pole from the centre of the map. Exaggeration on others is progressively greater away from the central meridian and especially equatorwards of the centre and towards the edges	Increases away from the centre	Beyond about 30° of the centre, shape is increasingly distorted because of increasing meridian and parallel scales and increasing obliquity of meridian and parallel	Areas are progressively exaggerated away from the centre. They are reasonable within 30° of the centre	As in 11 and 12 above: (*a*) Direction from the centre of the map is correct. (*b*) Any straight line drawn on the map is a great circle	These properties, combined with reasonable shape and area within about 30° of the centre, give rise to good general-purpose and navigational maps. Often used in conjunction with Mercator for navigation. Because centre can be placed anywhere, has a very wide application

PROJECTION	SHAPE OF PARALLELS	SHAPE OF MERIDIANS, etc.	INTER-SECTION OF MERIDIANS AND PARALLELS	SCALE ALONG MERIDIANS	SCALE ALONG PARALLELS	REPRE-SENTATION OF SHAPE	REPRE-SENTATION OF AREA	OTHER PROPERTIES	USES
14. POLAR ZENITHAL EQUIDISTANT	Concentric circles with pole as centre. Drawn their true distance apart	Straight lines radiating from the pole. Drawn their true *angular* distance apart	Right angle	Correct	Progressively exaggerated away from the pole; but not as rapidly as in Polar Zenithal Gnomonic; 1.2% at 75° latitude, 4.5% at 60° latitude and 11% at 45° latitude	Not orthomorphic, but reasonable shape within 30° of the pole. Equatorwards of this, progressive E.–W. elongation of land-masses	Areas of land-masses exaggerated progressively polewards, but not so rapidly as in Polar Zenithal Gnomonic	(*a*) Direction and (*b*) Distance from the centre of the map are both correct. Only half the globe can be shown	General-purpose maps of Arctic areas. Polar exploration and, to some extent, polar navigation
15. EQUATORIAL ZENITHAL EQUIDISTANT	Equator is a straight line of correct length. All other parallels are arcs of circles of progressively smaller radius polewards	Central meridian is straight line of correct length. All other meridians are arcs of circles of progressively smaller radius away from the centre. Bounding meridians are semicircles	Equator cuts all meridians at right angles. Central meridian cuts all parallels at right angles. Intersection progressively more acute polewards and away from the central meridian	The central meridian is of correct scale. All others are progressively exaggerated away from the central meridian. The bounding meridians are 57% exaggerated. Scale along each meridian is uniform throughout their length, i.e. parallels are everywhere equidistant from each other	Equator correct; all others either slightly reduced or exaggerated. But all are equally (if not correctly) divided. Hence, all meridians are equidistant from each other	Progressively badly distorted towards the margins of the map. Land-masses markedly elongated N.–S. towards the margins; but, near the centre, shape is fairly good	Areas progressively exaggerated towards the margins of the map mainly because of N.–S. elongation of land-masses, but *not* so rapidly as in Equatorial Zenithal Gnomonic	(*a*) Direction and (*b*) Distance from the centre of the map are both correct. Only half the globe can be shown	Navigation, especially aerial navigation from any airfield located; on the equator but airports on or near the equator are few
16. OBLIQUE ZENITHAL EQUIDISTANT	Curved lines, not arcs of circles. Slightly elliptical. *Not* parallel to each other	Composite curves converging on the pole	Central meridians cut the parallels at right angles. Intersection is more oblique towards the margins of the projection	Correct only along the central meridian. All others progressively exaggerated towards the margins. The	Each parallel is exaggerated in length equatorwards and the scale along each increases progressively towards the	There is a progressive distortion of shape towards the periphery. Near the periphery there is a 57% elongation of land-	Away from the centre of the map in all directions, there is progressive exaggeration of area, e.g. on the projection centred on	(*a*) Direction and (*b*) Distance from the centre of the map are both correct. Only half the globe can be shown	A most valuable projection for aerial navigation or for showing air, rail and/or sea routes from one point which is the centre of the projection.

				not equally spaced on the equatorial case			... and the part of South America which can be shown are very much exaggerated in area		... two cases that the centre of the map can be anywhere on the globe
17. EQUATORIAL ZENITHAL EQUAL-AREA	The equator is a straight line about 10% too short. All other parallels are arcs of circles of progressively smaller radius polewards	The central meridian is a straight line also about 10% too short. All other meridians are arcs of circles of progressively smaller radius away from centre	The equator cuts all meridians at right angles. The central meridian cuts parallels at right angles. Intersection is progressively more acute polewards and away from the central meridian	Length of the central meridian is diminished by 10%. Scale along it is diminished throughout, but more so towards the poles. One meridian on either side of the central is correct, but towards edges, they become progressively too long; the bounding meridian is 41% exaggerated. But within 40° of the centre, meridian scale diminution is very small	Length of the equator is diminished by 10%. Scale along it is diminished throughout, but more so towards the periphery. Other parallels are too short but not greatly diminished	Shape of land-masses is badly distorted towards the margins. Land-masses are compressed longitudinally, elongated latitudinally and 'pulled out of vertical' by the obliquity of the meridians to the parallels	Equal-area	True direction from the centre of the map is maintained by a straight line. Only half the globe can be shown	Of little value for a complete hemisphere because of peripheral distortion of shape. Within 40° of the centre, distortion of shape is very small. Africa, or South-east Asia (with Australia) or Central America and the Caribbean are well represented on it
18. OBLIQUE ZENITHAL EQUAL-AREA	Parallels are composite curves, not arcs of circles. *Not* parallel to each other	Meridians are composite curves, converging on the pole	Central meridians cuts the parallels at right angles. Intersection is more oblique towards the margins of projection	Length of the central meridian is diminished by 10%. Scale along it is diminished progressively away from the centre. Other meridians are longer than the central meridian and, therefore, diminished to a lesser extent	Length of each parallel is exaggerated away from the centre to compensate for diminution of the meridians	Land-masses are increasingly compressed latitudinally and correspondingly elongated longitudinally. Within about 40° of the centre, shape is very well preserved	Equal-area	True direction from the centre of the map is maintained by a straight line. Only half the globe can be shown	Of little value for a complete hemisphere because of latitudinal compression and longitudinal elongation towards the periphery of the map. But very suitable for a continent in one hemisphere limited to 40° or so of the centre, e.g. North America, Europe, even Asia.

PART II

THE CYLINDRICAL GROUP OF PROJECTIONS

Cylindrical projections are based upon the conception of a cylinder circumscribing the globe (Fig. 10) and touching it along the equator (though in the case of the Transverse Mercator it can touch the globe along any meridian). Meridians and parallels are transferred from the globe on to the circumscribing cylinder. When unrolled, the cylinder becomes a rectangle whose length is the same as that of the equator, i.e. $2\pi r$, where r is the radius of the globe. In all members of this group, the parallels are straight lines of the same length as the equator while the meridians are *also* parallel straight lines their correct distance apart at the equator and cutting the parallels of latitude at right angles. There are three *main* cylindrical projections: (*a*) the Simple Cylindrical, (*b*) the Cylindrical Equal-area (often called Lambert's) and (*c*) Mercator's or the Cylindrical Orthomorphic.

The only difference between these three is the spacing of the parallels. In the Simple Cylindrical they are drawn their correct distance apart, i.e. they are spaced exactly as they are on the globe. In Lambert's Cylindrical Equal-area, the parallels become progressively closer together polewards, i.e. the spacing of the parallels is so arranged that at any point in latitude the *reduction* in scale along the meridian *exactly compensates* for the *exaggeration* of the scale along the parallel. This makes the projection *equal-area* because at any point the meridian scale is as much too small as the parallel scale is too big, i.e. the meridian and parallel scales are compensatory. (See page 3.) In Mercator, the parallels become progressively further apart polewards. The spacing of the parallels in this case is arranged in such a way that at any point on the projection the scale along the meridian is the *same* as the scale along the parallel. This makes the projection *orthomorphic* because already the meridians and parallels are set at right angles to one another. In this way, the two requisites of orthomorphism are satisfied. (See page 9.)

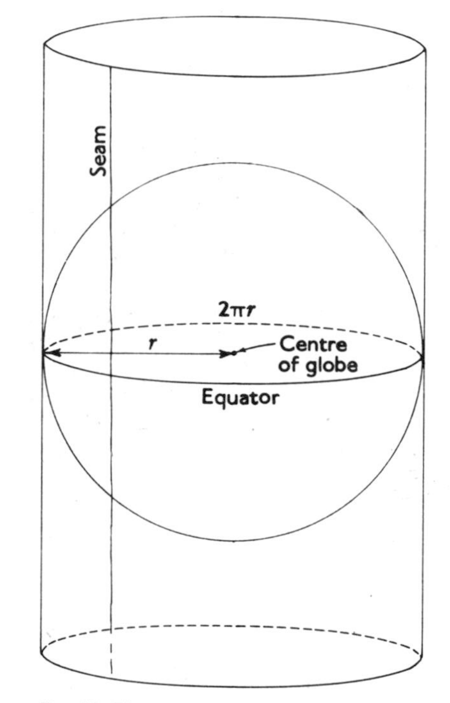

Fig. 10 The principle of cylindrical projection.

3 The Simple Cylindrical Projection (Fig. 11)

Construction

(*a*) The equator is drawn its true length, i.e. $2\pi r$, and is subdivided correctly. If the meridians are drawn at intervals of 15°, each subdivision of the equator is

$\frac{2\pi r}{360}\times 15^\circ = \frac{2\pi r}{24}$ or $\frac{\pi r}{12}$ (See page 4.)

(*b*) Meridians are drawn at right angles to the equator through each point of subdivision.

(*c*) Parallels are then drawn as parallel straight lines their true distance apart, i.e. their curved or arc distance apart on a globe of radius r. If drawn at intervals of 15°, their distance apart is $\frac{\pi r}{180}\times 15^\circ = \frac{\pi r}{12}$

Because the intervals between meridians and parallels are both 15°, the resulting projection is a series of squares and is sometimes called the 'chess-board' or 'plate carrée'.

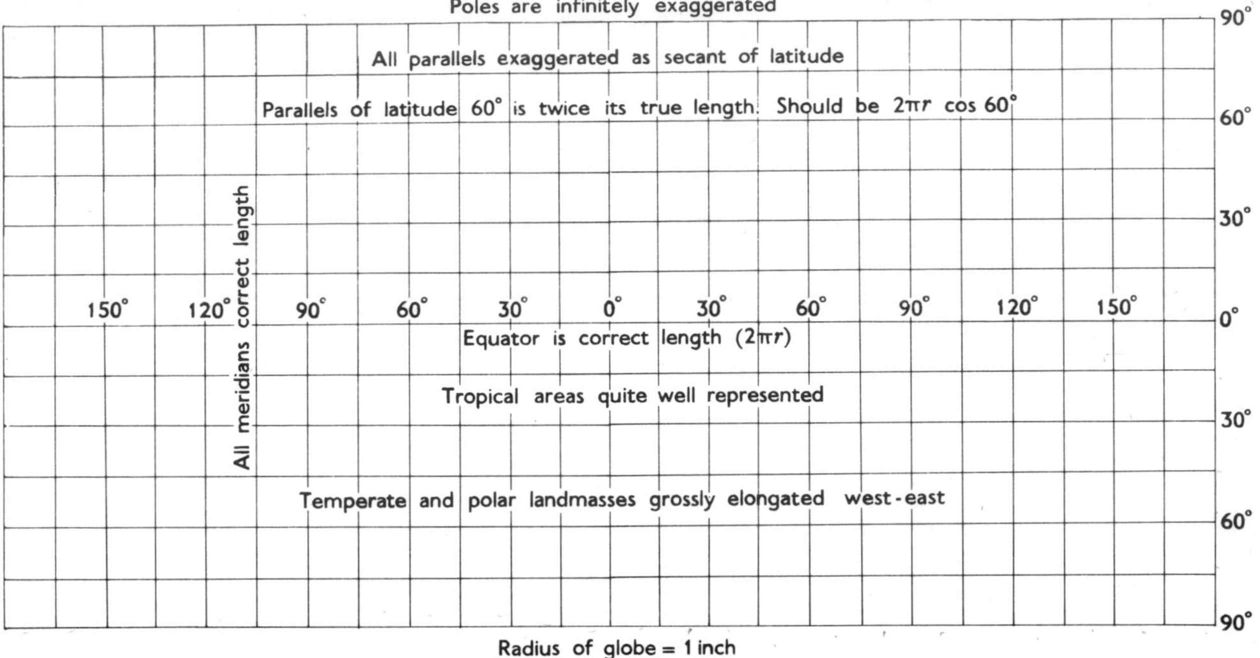

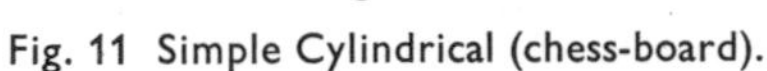
Fig. 11 Simple Cylindrical (chess-board).

On all cylindrical projections, because all parallels are drawn the same length as the equator ($2\pi r$) and the equator is drawn correct to scale, all other parallels are exaggerated in length. The amount of exaggeration of the length of each may be calculated as follows:

$$\frac{\text{Length of parallel ON PROJECTION}}{\text{Length of parallel ON GLOBE}} = \frac{2\pi r}{2\pi r \cos \text{lat.}}$$

$$= \frac{1}{\cos \text{lat.}} = \text{secant lat.}$$

The exaggeration of the length of any parallel of latitude may, therefore, be discovered by consulting a table of secants. From such a table, it will be noticed that the parallel scale is progressively exaggerated polewards, so that while the equator is the correct length, parallel 60°N. is twice its true length and the pole infinitely exaggerated, i.e. it *is* of *finite* length whereas it should be a point. Exaggeration of scale at the pole $\frac{2\pi r}{0} = \infty$

The scale is correct along all meridians and along the equator. The projection is *not orthomorphic* because, although the meridians and parallels intersect at right angles, the scale along the meridian at any point is not the *same* as the scale along the parallel at that point. For example, at 60°N. the scale along the parallel is twice what it should be, but the scale along the meridian at 60°N. is correct. Clearly, therefore, the latitudinal extent of landmasses is correct to scale but their longitudinal (east–west) extent becomes progressively exaggerated polewards. Equally clearly, because the meridian and parallel scales at any point are neither (*a*) both correct, nor (*b*) compensatory, the projection cannot be equal-area.

Uses

For world maps, therefore, the projection is of little value because of progressive areal exaggeration and distortion of shape polewards. However, the areal exaggeration at latitude 30° is about 15·5% and at the Tropics only 9% so that within the Tropics neither area nor shape is *very* badly distorted. Despite this, it is seldom used even for representation of tropical areas because other projections, equally easy to draw are better, for example, the Cylindrical Equal-area has only a slight distortion of shape within the Tropics, while that associated with Sanson–Flamsteed's and Mollweide's (both equal-area) in those regions is even less.

4 The Cylindrical Equal-Area Projection (Lambert's)

The Cylindrical Equal-area (Fig. 12), like most other cylindrical projections, is not strictly projected but in so far as the planes of each parallel of latitude are *extended* to the circumscribing cylinder, the result is the same as would be expected if the parallels had been 'projected' from infinity. We may, therefore, look upon the Cylindrical Equal-area as a kind of 'Cylindrical Orthographic'.

The parallels of latitude are represented as parallel straight lines all the same length as the equator which is drawn correct to scale. The meridians are also parallel straight lines at right angles to them. The parallels of latitude become progressively closer together polewards and are so spaced that the area between any two of them is made equal to the area of the corresponding zone on the globe.

The meridians, being drawn through points of correct subdivision of the equator, are equidistantly spaced.

To achieve the equal-area property, the distance of any parallel from the equator is r sin latitude where r is the radius of the globe. (For explanation and proof of this, see end of chapter.)

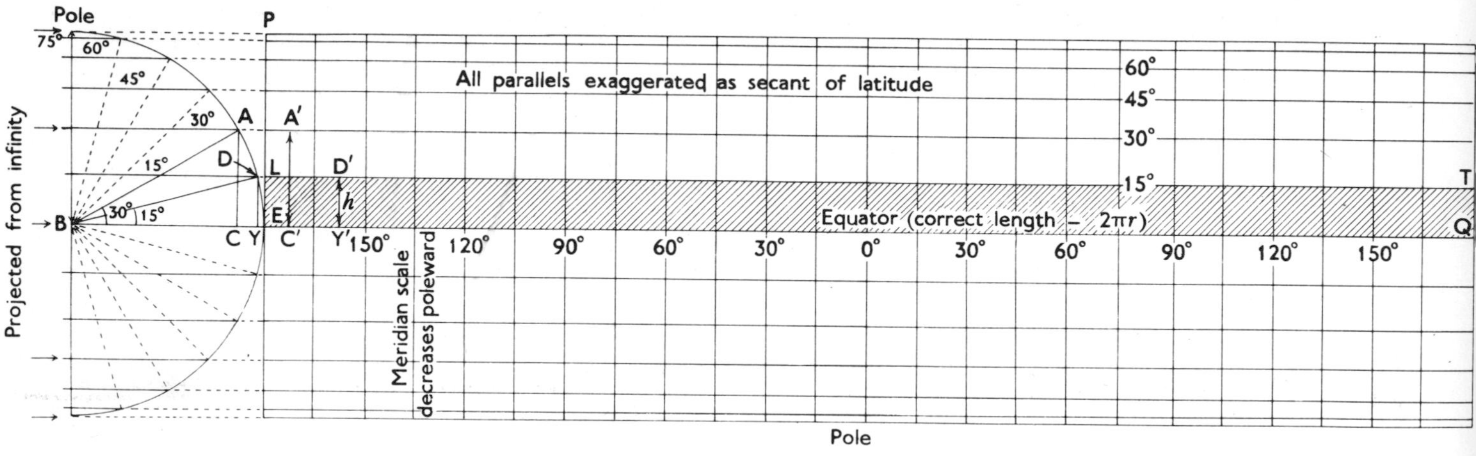

Fig. 12 The Cylindrical Equal-area (Orthographic).

Calculation of a Simple Example

(a) Length of equator $= 2\pi r$.

(b) Subdivision of equator (meridians at 15° intervals)

$$= \frac{2\pi r}{360} \times 15 = \frac{2\pi r}{24}.$$

(c) Distance of each parallel from equator $= r$ sin latitude:
15° (North and South): r sin 15°
30° (North and South): r sin 30°, etc.

Characteristics, Properties and Uses

(1) The scale along the equator is correct and, in common with all other cylindrical projections, the other parallels are progressively exaggerated in length polewards as the secant of the latitude.

(2) The scale along the meridians is progressively diminished polewards, as the parallels become closer together.

(3) Meridians and parallels intersect at right angles.

(4) Because the meridian and parallel scales are compensatory, i.e. because at any point the reduction of the meridian scale exactly compensates for the exaggeration of the parallel scale, the projection is equal-area (see end of chapter).

(5) Clearly, the projection cannot also be orthomorphic. Although the meridians and parallels *do* intersect at right angles, the scale along the meridian at any point is too small while the scale along the parallel at the same point is correspondingly too large.

The shape of land-masses is, in fact, badly distorted polewards of about 45° latitude. They are elongated east–west because of the exaggeration of the parallel scale and are compressed north–south because of the diminution of the meridian scale polewards. At 60° latitude, for example, the east–west extent of the land-masses would be doubled and they would be correspondingly compressed in latitude; at 75° latitude, the east–west elongation is nearly fourfold (in fact, 3·86 times). Because of the greater extent of land-masses, the distortion of shape is especially noticeable in the northern hemisphere—Eurasia and Canada being badly affected.

As a projection for a map of the world it is, therefore, most unsatisfactory; but shape is reasonably well preserved in the Tropics. *At* the Tropics, the east–west stretching is only 9% and the scale along the meridian at 23½° latitude is diminished by only 9%. (Note that the secant of 23½° is 1·0904.) This is not great.

The correct representation of area and reasonable representation of the shape of land-masses in tropical latitudes makes it suitable for all forms of distribution of these areas.

Calculation

To show that the projection is equal-area (See Fig. 12)

(1) The area of a sphere $= 4\pi r^2$

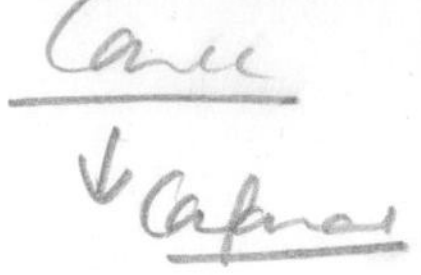

The area of a hemisphere ON THE GLOBE $= 2\pi r^2$
The area of a hemisphere ON THE PROJECTION =
Length of equator × Perpendicular distance between equator and pole
$= EQ \times EP = 2\pi r \times r \qquad = 2\pi r^2$
Clearly, therefore,
the area of a hemisphere ON THE GLOBE
= the area of a hemisphere ON THE PROJECTION
But to be truly equal-area, *all parts* of the projection must be equal in area to their counterparts on the globe.
(2) The area of a zone on the globe (See Fig. 13) =

$2\pi r$	×	h
(The length of the equator)		(The perpendicular distance between the bounding parallels.)

This is precisely the area of the zone represented by the rectangle *LTQE* (shaded on Fig. 12).

Examples

(*a*) Area of zone between equator and 15°N. (Fig. 13)
$= 2\pi r \times D'Y'$
$= 2\pi r \times r \sin 15° = 2\pi r^2 \sin 15°$
(*b*) Area of zone between 15°N. and 30°N. $= 2\pi r \times A'C'$
$= 2\pi r\,(r \sin 30 - r \sin 15) = 2\pi r^2\,(\sin 30 - \sin 15)$

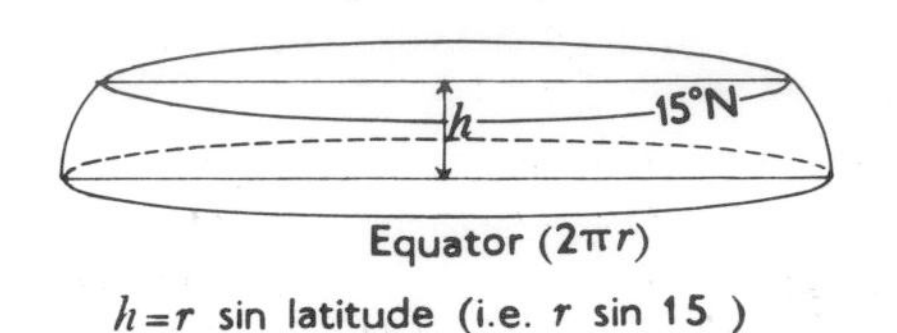

Fig. 13 The area of a zone ($2\pi r.h$).

To show that the distance of any parallel from the equator = r sin latitude (see Fig. 12)

In △ *ABC*, angle *ABC* = angle of latitude 30° and
$AB = r$, the radius of the globe.

$$\frac{AC}{AB} = \sin \angle ABC$$

Therefore, $AC = AB \sin \angle ABC$
$= r \sin \text{latitude } 30° = A'C'$
Similarly, $DY = D'Y' = r \sin 15°$
The distance of the pole from the equator
$= r \sin \text{latitude} = r \sin 90°$
$= r \times 1 = r = PE.$

5 Mercator's Projection (Fig. 14)

Mercator (1512–94) was a Dutch geographer who prepared a chart for which he devised a novel projection which enabled seamen during the Age of Discovery accurately to determine compass bearings between one place and another. The Mercator projection became, and remains, the standard projection for navigation charts. It was so widely accepted that it was often used for purposes for which it was never intended, and until about the end of the nineteenth century, maps of the world based upon it were looked upon as the only correct maps. Because users of it were not fully aware of its properties, it gave rise to many misapprehensions about the relative size and disposition of land-masses on the earth.

As the projection falls into the cylindrical group—it is the Cylindrical Orthomorphic—it has a number of characteristics which are common to all cylindrical projections. First, meridians and parallels are straight lines set at right angles. Secondly, the equator is the correct length, but as all other parallels are the same length as the equator, they are progressively exaggerated in length polewards. The amount of exaggeration of the length of any parallel may be calculated as follows:

$$\frac{\text{Length of parallel ON PROJECTION}}{\text{Length of parallel ON GLOBE}} = \frac{2\pi r}{2\pi r \cos \text{lat.}} = \frac{1}{\cos \text{lat.}} = \text{secant lat.}$$

So the exaggeration of the length of any parallel may be obtained by consulting a table of secants:

	Secant	*Percentage Exaggeration*
0° (the equator)	1·0000	Nil
15°	1·0353	3·5
30°	1·1547	15·47
45°	1·4142	41·42
60°	2·0000	100
75°	3·8637	286·37
90°	∞	∞

It is seen that there is a very considerable east–west stretching especially polewards of 30° latitude. Because the meridians and parallels intersect at right angles, the projection can be made orthomorphic by the apparently simple device of balancing the east–west stretching by an equal north–south stretching at each point on the projection.

Because 60°N., for example, is twice its correct length, the scale along the meridian at 60°N. must also be made twice its correct length. Similarly, 80°N. is exaggerated 5·759 times, so the meridian at 80°N. must be stretched to the same extent.

Clearly, the pole cannot be shown on this projection. On the globe, the pole is a point, but on any cylindrical projection, it would be represented as a straight line of the same length as the equator and would, therefore, be infinitely exaggerated. To maintain the property of orthomorphism, the scale along the meridian at the pole would also have to be infinitely exaggerated and so the pole would be infinitely distant from the equator. One would require a sheet of paper infinitely long in order to represent it!

Representation of Area (*Fig.* 15)

Because the parallels are stretched progressively poleward as the secant of the latitude and because at any point on the projection this east–west stretching is balanced by an equal north–south stretching, area must be exaggerated at any point as the square of the secant of the latitude. For example, because at 60°N. both parallel and meridian scales are doubled (that is, both exaggerated at the secant of 60°, which is 2·0), area is quadrupled. At 70° latitude, areal exaggeration is secant 70° squared, i.e. $(2{\cdot}9238)^2$—nearly ninefold, at 80° latitude, $(5{\cdot}759)^2$ or over thirty-two fold. (See Fig. 15.) Obviously, the areas of land-masses in high latitudes are greatly exaggerated; Canada, the U.S.S.R., Greenland and Alaska appear to be enormous. South America is about nine times the size of Greenland, but on Mercator Greenland is the bigger.

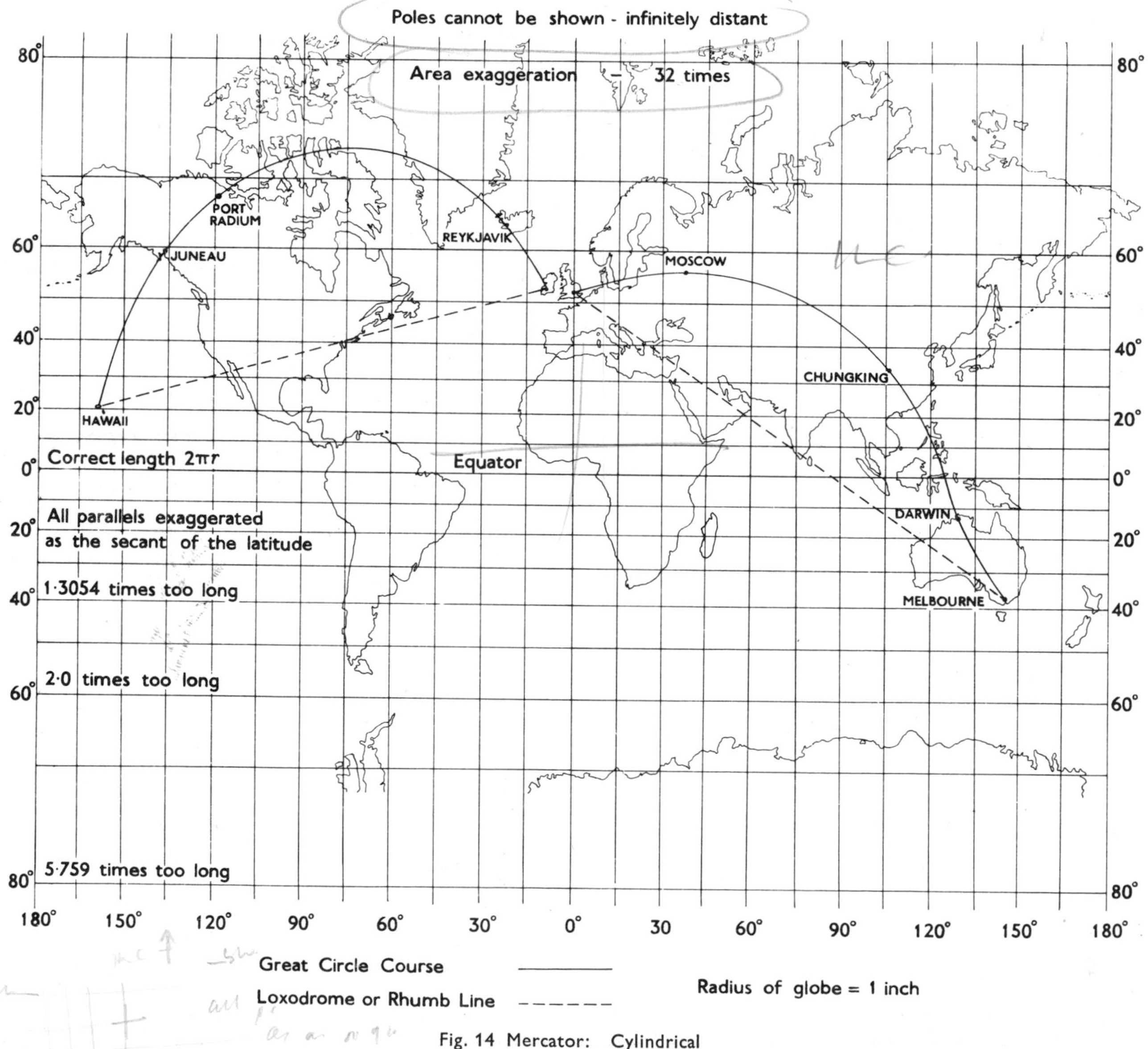

Fig. 14 Mercator: Cylindrical Orthomorphic.

Representation of Shape on Mercator

Mercator's projection shows clearly what is meant by 'orthomorphism'. It does not simply mean 'correct shape'. The shape of North America demonstrates this; taken as a whole, it is 'top-heavy' because the exaggeration of area in arctic latitudes is about 6·6 times (or 660%) while that in Florida and along the Gulf Coast is between 10 and 13%. The shape of the sub-continent as a whole cannot, therefore, be correct even though the shape of each small area, strictly speaking, each *infinitely small* area, is correct.

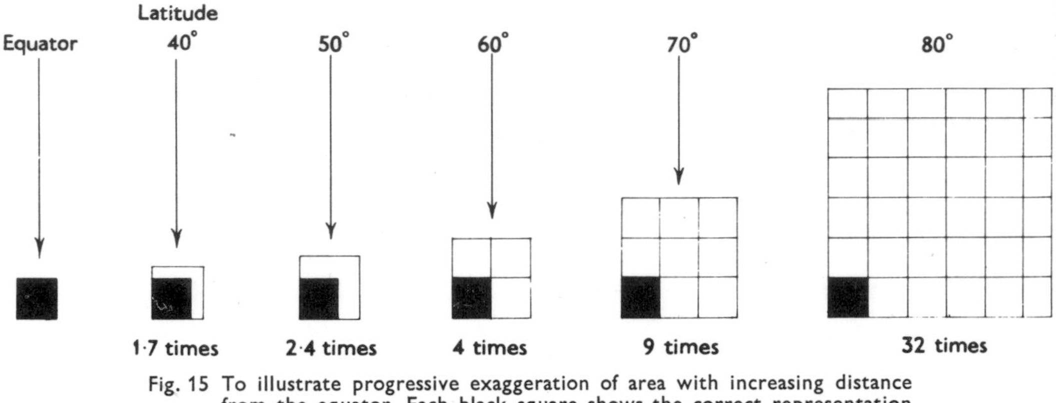

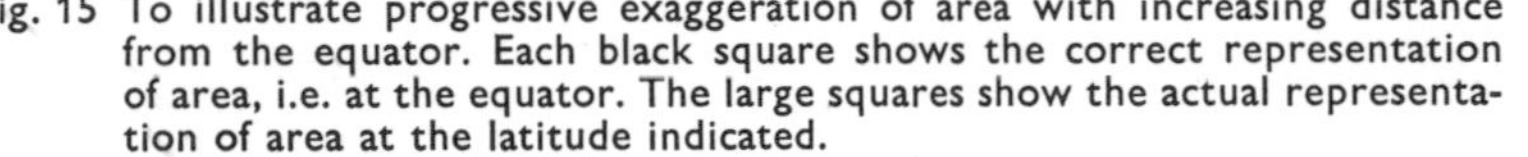

Fig. 15 To illustrate progressive exaggeration of area with increasing distance from the equator. Each black square shows the correct representation of area, i.e. at the equator. The large squares show the actual representation of area at the latitude indicated.

Direction on Mercator

A straight line on Mercator is called a 'loxodrome' or 'rhumb-line'. (Fig. 14.) It maintains a constant direction and is, accordingly, a line of constant bearing. It maintains its accuracy of direction simply because Mercator is orthomorphic; that is, at any point on the projection, any east–west stretching of the parallel is matched by a corresponding stretching of the meridian, while at the same time the right-angled intersection of meridian and parallel is maintained. Only under these conditions can accuracy and constancy of direction be achieved by a straight line. So the compass bearing to be followed in order to travel between one place and another can readily be determined by first joining the two places on a Mercator chart by a straight line and then measuring the angle between it and the meridian or true-north line. But the straight line would not represent the shortest distance between the two points. This would be along the great circle which passes through both points.

If a piece of string is held tightly on the globe between London and Melbourne, it would with reasonable accuracy delineate the Great Circle route which would pass through Hamburg, Moscow, Karaganda, Chungking, the island of Celebes and Darwin. Similarly, the Great Circle route between Hawaii and Shannon crosses Juneau (Alaska), Port Radium on the eastern shores of Great Bear Lake, northern Baffin Island and Reykjavik. Plotted on Mercator, these Great Circle courses would appear as curved lines, convex polewards. (See Fig. 14.) They are difficult to plot and published tables are usually used for the purpose.[1] Even when plotted on Mercator, it cannot be navigated by compass because its compass direction is constantly changing. It is not possible to move along it without changing direction. The best that can be achieved is a very close approximation to the Great Circle course by setting off loxodromes or rhumb-lines as a series of 'legs' along the plotted Great Circle and, by changing direction at the beginning of each 'leg' (points 2 to 4 on Fig. 16). Clearly, the greater the number of 'legs', the more closely would the loxodromic route approximate to the Great Circle or shortest route.

[1] Alternatively, the Great Circle Course could be drawn on the Zenithal Gnomonic (on which it appears as a straight line between the two terminal points) and then transferred to the Mercator projection.

Uses

Mercator should never be used for maps of land-masses because of the great areal exaggeration in high latitudes. It is especially suitable for sea and air navigation and for any other purpose for which correct direction is necessary, for example, the showing of ocean currents or wind direction. Distributions of climatic regions or natural vegetation should not, however, be plotted on the same map as these require an equal-area graticule for making regional comparisons of area. Mercator would give a very false impression of the relative areas covered by coniferous and equatorial forests or by hot and cold deserts.

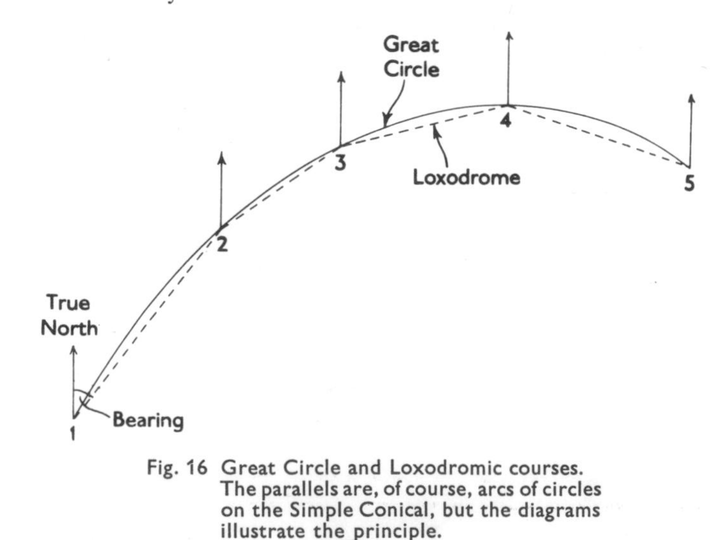

Fig. 16 Great Circle and Loxodromic courses. The parallels are, of course, arcs of circles on the Simple Conical, but the diagrams illustrate the principle.

6 Gall's Projection (Fig. 17)

In Gall's projection, it is assumed that:
(1) the cylinder cuts the globe along parallels 45°N. and 45°S. and
(2) the parallels are projected stereographically, i.e. from one end of the equator.

Meridians and parallels are straight lines intersecting at right angles as in all cylindrical projections. Similarly, the parallels are all the same length—in this case $2\pi r \cos 45$ which reduces them to about 7/10ths (0·7071 times) the length of parallels on other cylindrical projections. The only parallels which are of correct length are parallels 45°N. and 45°S. All parallels equatorwards of these become progressively too small, the equator being reduced to about 7/10ths of its length on the globe. However, polewards of 45° latitude, there is a fairly rapid exaggeration of the parallel scale; 60° is about 1·4 times too long, 75° nearly $3\frac{1}{2}$ times too long and the pole infinitely too long. The reduction or exaggeration, as the case may be, of the scale along the parallels may be calculated as follows:

$$\frac{\text{Length of parallel ON PROJECTION}}{\text{Length of parallel ON GLOBE}} = \frac{2\pi r \cos 45}{2\pi r \cos \text{lat.}} = \frac{\cos 45}{\cos \text{lat.}}$$

The scale along the meridian increases progressively polewards, but it is correct only in the neighbourhood of latitude 45°. At all points along the meridians between the equator and 45°, it is too small. This is shown on Fig. 17 by the fact that *A'R* the distance of parallel 10° from the equator is smaller than *AQ* which is its arc distance from the equator on the globe; similarly, *A'B'* is smaller than *AB* and *B'C* smaller than *BC*—but to a progressively smaller extent. However, north of 45° latitude, the meridian scale is progressively exaggerated polewards. Fig. 17 shows that *CD* is greater than *CD'*, *DE* greater than *D'E'* and *EF* greater then *E'F'*.

Equatorwards of latitude 45° both the meridian and parallel scales are too small, but at no point are they the *same*. Polewards of latitude 45° both the meridian and the parallel scales are too great, but everywhere the parallel scale is much greater than that along the meridian. It is clear, therefore, that the projection is neither equal-area nor orthomorphic. To be equal-area, the meridian and parallel scales must be either both correct or compensatory; nowhere on the projection is this so. To be orthomorphic, not only must the meridians and parallels intersect at right angles, but everywhere the meridian and parallel scales must be exaggerated or diminished in the same ratio. Again, as has been shown, nowhere on the projection is this so.

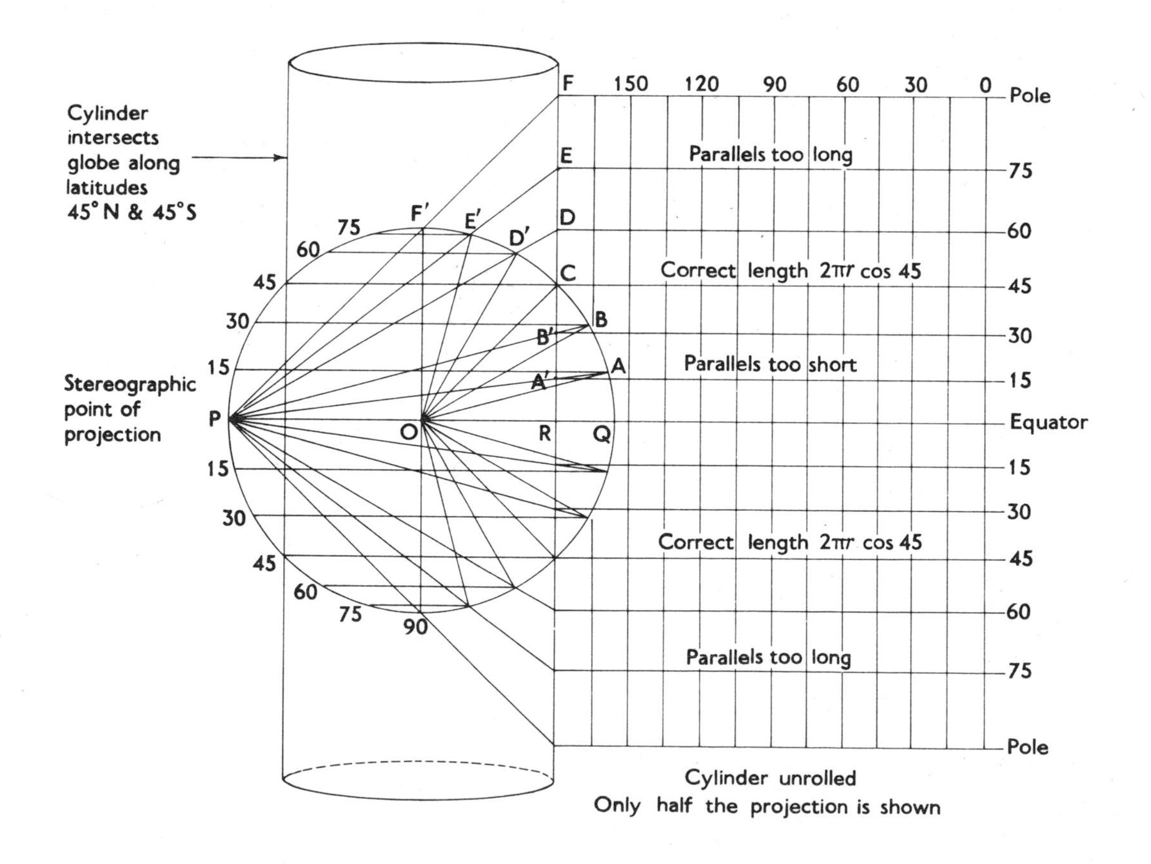

Fig. 17 Gall's projection.

Areas of land-masses are progressively diminished equatorwards of 45° latitude; but polewards of 45° they are progressively enlarged, *partly* because of the exaggeration of the meridian scale, but *mainly* because of the elongation of the parallels. For this same reason the land-masses in arctic latitudes appear to be compressed. In fact, there is no compression; on the contrary, they are *stretched* north–south but appear to be compressed only because they are *more* than proportionately stretched from east to west.

The northern continents, especially North America, show a 'top-heaviness' which is reminiscent of Mercator. Notice particularly the east–west elongation of Alaska, the mainland of northern Canada, Baffin Island, Greenland, northern Eurasia and, especially, the peninsula of north-eastern Siberia—all to the north of latitude 60° beyond which the elongation of the parallels is greatest.

Gall's is a projection which attempts to effect a compromise between accuracy of shape and area. Only in arctic areas is shape noticeably bad and area noticeably disproportionate. It has been used, therefore, in atlases for showing distributions as in Johnston's *Advanced Modern School Atlas*, but in others, for example, in Philips' *Modern School Atlas*, an equal-area projection, the interrupted Mollweide is used.

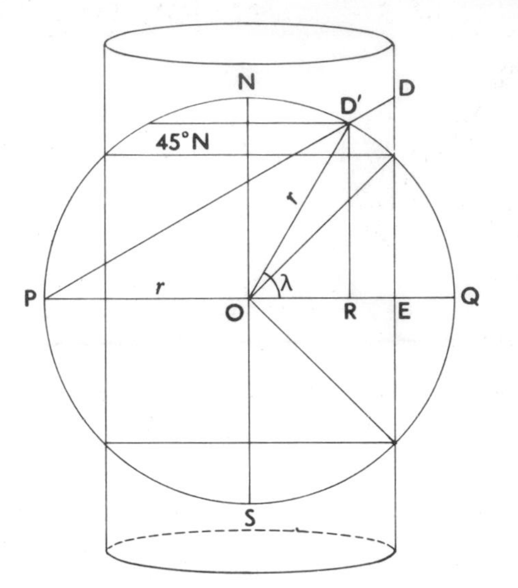

Fig. 18 To find distance of the parallels from the equator on Gall's.

Calculation

Radius of the globe:—1 inch.

(1) Length of all parallels $= 2\pi r \cos 45 = (2 \times 3{\cdot}142 \times 0{\cdot}7071)$ inches.

(2) Spacing of the meridians (at 15° intervals)

$$= \frac{2\pi r \cos 45}{360} \times 15 = \frac{2 \times 3{\cdot}142 \times 0{\cdot}7071}{24} = \frac{4{\cdot}4434}{24} = 0{\cdot}185 \text{ ins.}$$

(3) Distance of parallels from the equator

$$= \frac{r \sin \text{latitude}\,(1 + \cos 45)}{1 + \cos \text{latitude}}$$

(*Proof.* (*see Fig.* 18) *To find value of ED*)

(*a*) In triangle $D'OR$, $\frac{D'R}{D'O} = \sin\lambda$; therefore, $D'R = D'O \sin\lambda = r \sin\lambda$.

(*b*) $OR = r \cos\lambda$; therefore, $PR = r + r\cos\lambda$.

(*c*) $OE = r \cos 45$; therefore, $PE = r + r \cos 45$.

(*d*) Triangles PRD' and PED are similar;

therefore, $\frac{ED}{D'R} = \frac{PE}{PR}$ i.e. $ED = \frac{D'R \times PE}{PR}$

Substitute values as above:

$$ED = \frac{D'R \times PE}{PR} = \frac{r \sin\lambda\,(r + r\cos 45)}{r + r\cos\lambda}$$

PART III

THE CONICAL GROUP OF PROJECTIONS

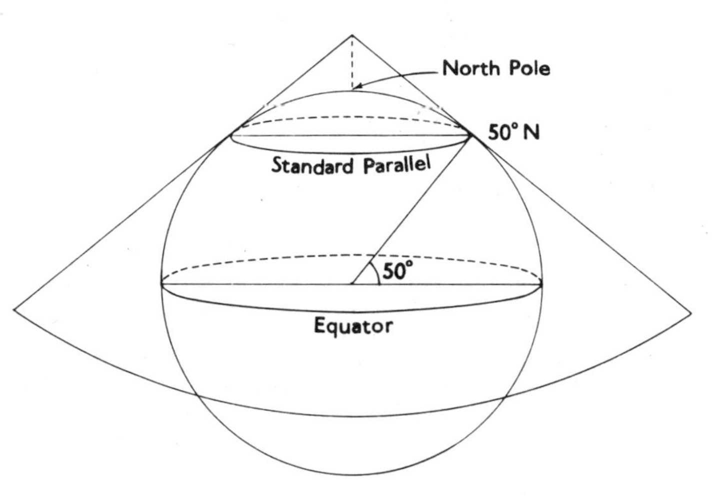

Fig. 19 The principle of conical projection.

In the construction of most projections in this group, a cone is assumed to be placed over the globe so that its apex is vertically above the pole, i.e. on the axis of the globe produced. The cone, therefore, touches the globe along a single parallel of latitude. This parallel is known as the *standard parallel.* (Fig. 19.)

The simpler forms of conical projection, such as the Conical with One *and* the Conical with Two Standard Parallels, all have meridians which radiate as straight lines from a point on the central meridian produced, and parallels which are concentric circles.

7 The Conical Projection with One Standard Parallel

Stages in Construction (see Fig. 26(*a*) *and* (*b*))

(1) The central meridian is drawn as a vertical straight line through the centre of the area to be mapped.
(2) The standard parallel is then chosen; this should also pass approximately through the centre of the area to be mapped. It is drawn as an arc of a circle with centre A and radius AL. (Fig. 21.) The value of AL is r cot latitude (i.e. or r tan colatitude). Those readers interested in the proof of this should refer to 'Calculation' at the end of the chapter.
(3) The standard parallel is then subdivided correctly, the subdivisions being marked off from the central meridian on either side. If meridians are drawn at intervals of 10° the subdivisions are:

$$\frac{\textit{Length of standard parallel on globe}}{360} \times 10$$
$$= \frac{2\pi r \text{ cos latitude of standard parallel}}{36}$$

(4) The other parallels are then inserted as concentric circles their true distance apart by subdivision of the central meridian, starting from the standard parallel. If parallels are drawn at intervals of 10° the subdivision is:

$$\frac{\text{Length of central meridian on globe}}{180} \times 10^\circ$$
$$= \frac{\pi r}{180} \times 10 = \frac{\pi r}{18}$$

(5) The meridians are then drawn in as straight lines by joining the points of subdivision of the *standard parallel* to the centre of curvature of the parallels (i.e. the common centre of the concentric parallels of latitude).

Characteristics, Properties, Limitations and Uses (Fig. 22)

The meridians are straight lines radiating from the centre of curvature of the parallels—*not* the pole.

The parallels are arcs of concentric circles drawn their true distance apart.

Because the meridians are radii of the concentric circles, the meridians and parallels intersect at right angles.

The scale along the standard parallel is correct. All other parallels are too long and their exaggeration in length is

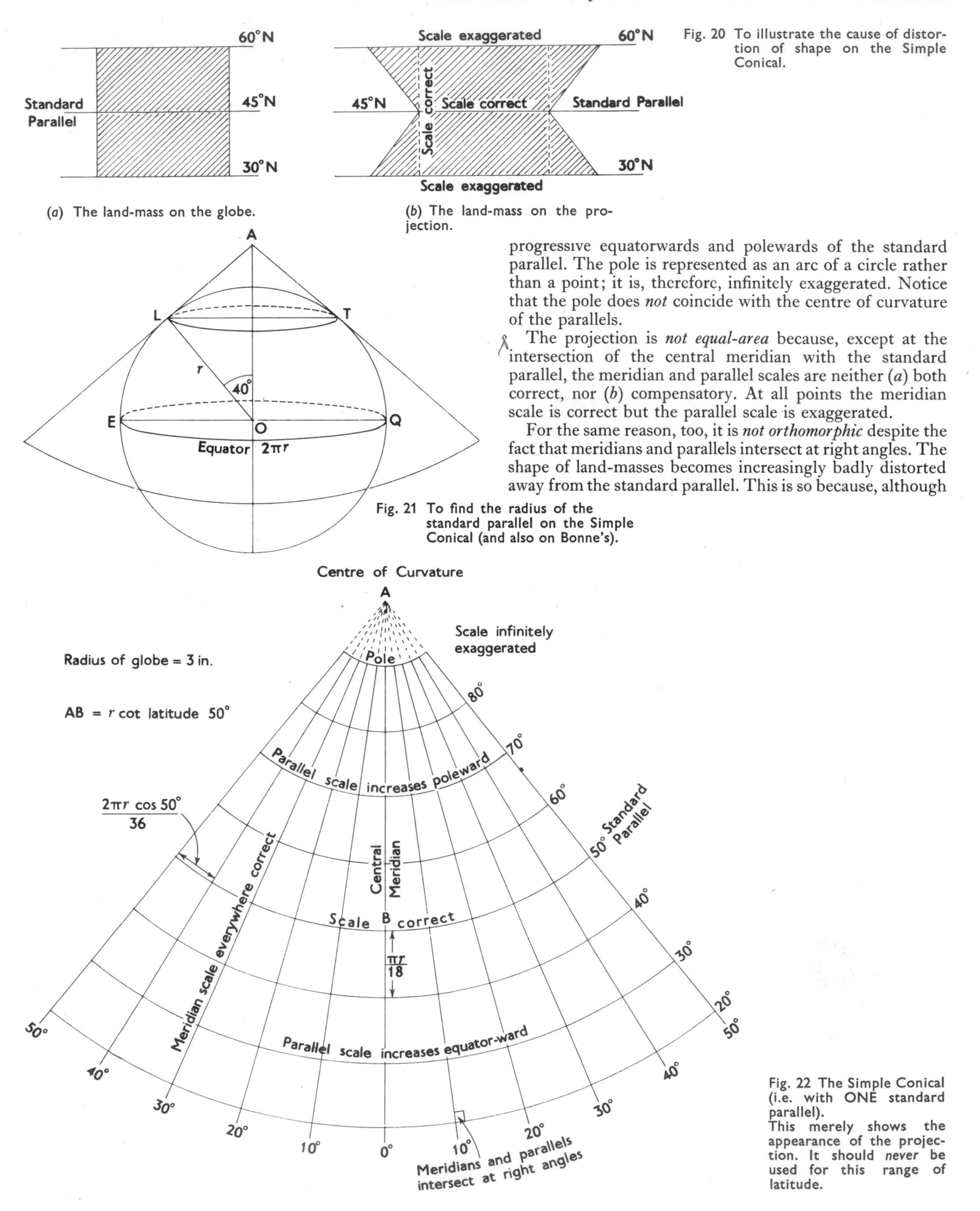

Fig. 20 To illustrate the cause of distortion of shape on the Simple Conical.

(*a*) The land-mass on the globe.

(*b*) The land-mass on the projection.

Fig. 21 To find the radius of the standard parallel on the Simple Conical (and also on Bonne's).

Fig. 22 The Simple Conical (i.e. with ONE standard parallel).
This merely shows the appearance of the projection. It should *never* be used for this range of latitude.

progressive equatorwards and polewards of the standard parallel. The pole is represented as an arc of a circle rather than a point; it is, therefore, infinitely exaggerated. Notice that the pole does *not* coincide with the centre of curvature of the parallels.

The projection is *not equal-area* because, except at the intersection of the central meridian with the standard parallel, the meridian and parallel scales are neither (*a*) both correct, nor (*b*) compensatory. At all points the meridian scale is correct but the parallel scale is exaggerated.

For the same reason, too, it is *not orthomorphic* despite the fact that meridians and parallels intersect at right angles. The shape of land-masses becomes increasingly badly distorted away from the standard parallel. This is so because, although the meridian scale is everywhere correct, the parallel scale becomes increasingly exaggerated away from the standard parallel, and this results in an increasing east–west elongation of land-masses away from the standard. This can be demonstrated by comparing the shape of a hypothetical square land-mass with its representation on the projection as in Fig. 20.

This progressive exaggeration of the *parallel* scale and its resulting east–west elongation of land-masses limits the use of the projection to countries which have a small extent in latitude. It is best used either for (*a*) countries such as the British Isles whose latitudinal extent is no more than 10° or other individual countries of Europe such as the Netherlands, Belgium, France—but not Scandinavia, or (*b*) for the representation of certain transcontinental railways such as the Trans Siberian, the Canadian Pacific or the Trans Andine, all of which roughly follow a particular parallel of latitude which can be chosen as the standard parallel, the only parallel which is correct to scale. 55°N. is roughly the median parallel for the Trans Siberian, 50°N. or 51°N. for the Canadian Pacific and 34°S. for the Trans Andine.

Even for these limited purposes, Bonne's or the Conical with Two Standard Parallels is preferred. Its use is, of course, confined to areas in one hemisphere.

Calculation

To find the radius of curvature of the standard parallel (i.e. AL in Fig. 21)

Angle LOE is the angle of latitude, 50°.
Angle LOA is the angle of colatitude, 40°.
In triangle ALO, angle $ALO = 90°$ and $LO = r$, the radius of the globe.

Therefore, $\frac{AL}{LO} = \tan \text{ angle } AOL = \tan 40°$ (i.e. tan colatitude)

$$AL = LO \tan \text{colatitude}$$
$$= r \tan \text{colatitude}$$

8 The Conical Projection with Two Standard Parallels

The Conical Projection with TWO Standard Parallels is a means of correcting the more serious defects of the Simple Conical projection, the Conical with ONE Standard Parallel. The latter has the great disadvantage that reasonable accuracy of the representation of land-masses is confined to a narrow zone on either side of the single standard parallel, which is the only parallel drawn to scale—all others being exaggerated in length progressively towards both the pole and the equator.

In the Conical Projection with Two Standard Parallels (Fig. 23), the error of the parallel scale is spread over a wider area by making TWO parallels, the two standards, correct to scale. Those between the two standards are slightly too short, while those on the poleward side of the One Standard Parallel and those on the equatorward side of the other increase progressively more rapidly in scale. The effect of this is that, although neither shape nor area is correct, the zone between the two Standard Parallels is reasonably well represented both in shape and in area, and so are the narrow belts beyond the two standards.

The radius on the projection of the standard parallel nearest the pole governs the curvature of the other parallels which are concentric and drawn their true distance apart. In Fig. 24, AB is the radius of the standard parallel nearest the pole. Those readers interested in determining mathematically its length should refer to 'Calculation' at the end of this chapter.

Characteristics and Properties

Meridians are all straight lines converging on a point on the central meridian produced, which is also the centre of curvature of the parallels.

Parallels are concentric circles drawn their correct distance apart. All meridians intersect the parallels at right angles because the meridians are the common radii of the concentric parallels; and all meridians are of correct length.

Of the parallels, only the two standards are of correct length. Those lying between them are slightly too short and those on the poleward side of the one and the equatorward side of the other are progressively exaggerated in length away from the standards. The pole is infinitely exaggerated because, as on the Conical with One Standard Parallel, it is represented by an arc of a circle rather than a point as it is on the globe.

The projection is not equal-area; except at the intersection of the standard parallels and the meridians, the meridian and parallel scales are nowhere either (*a*) both correct, or (*b*) compensatory.

Nor is the projection orthomorphic because although the meridians cut the parallels at right angles, the scales along both meridian and parallel are *not the same* at any selected point on the projection. Equality of meridian and parallel scales occurs *only* along the standard parallels where meridian and parallel scales are *both correct*. Elsewhere, while the meridian scale is by construction correct, the parallel scale is either too large or too small. However, there is not a very great diminution of the parallel scale between the standard parallels nor a very great exaggeration of it in the narrow zones outside the standard parallels. Therefore, while neither the shape nor the area of land-masses is correct, they are both reasonably well represented.

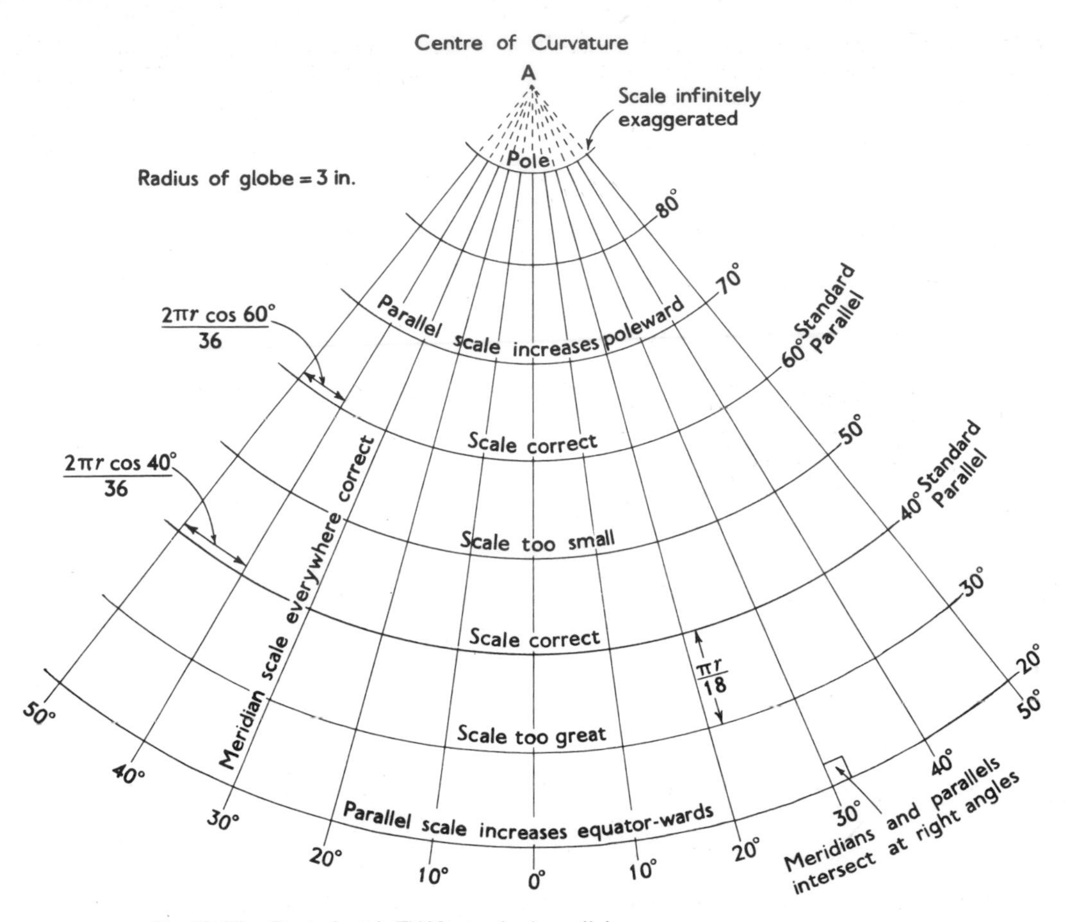

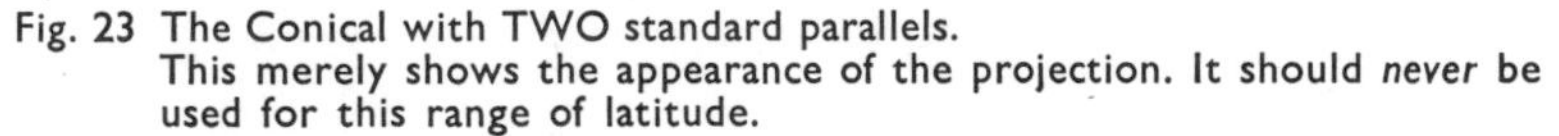
Fig. 23 The Conical with TWO standard parallels.
This merely shows the appearance of the projection. It should *never* be used for this range of latitude.

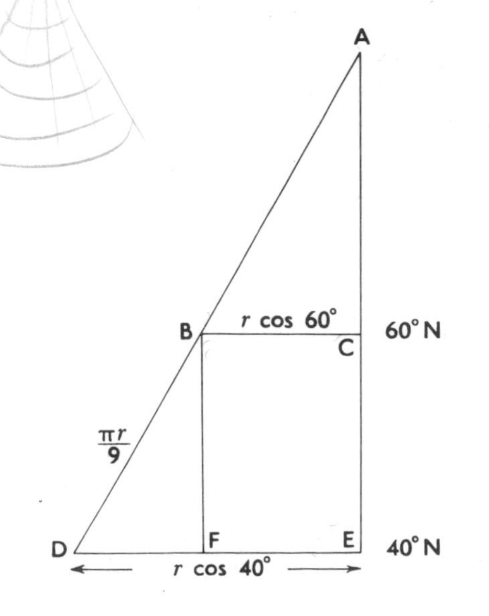

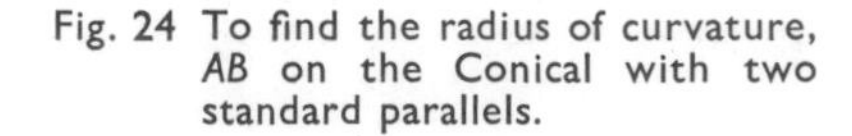
Fig. 24 To find the radius of curvature, *AB* on the Conical with two standard parallels.

Calculation

In Fig. 24, BC and DE are respectively the radii of the standard parallels 60°N. and 40°N. set at right angles to AE which represents the central meridian. They are so spaced that the slant distance, BD, between them is the correct distance between the two parallels on the globe, i.e. $\frac{\pi r}{180} \times 20° = \frac{\pi r}{9}$

AB is the radius of curvature of the standard parallel 60°N. which is required to be found.
Triangles ABC and BDF are similar, so that:

$\frac{AB}{BD} = \frac{BC}{DF}$; therefore, $AB = \frac{BD \times BC}{DF}$

But, $BD = \frac{\pi r}{9}$, $BC = r \cos 60°$ and $DF = r \cos 40 - r \cos 60$.

Substituting, therefore,

$$AB = \frac{\pi r/9 \times r \cos 60}{r(\cos 40 - \cos 60)} = \frac{\pi r/9 \times \cos 60}{\cos 40 - \cos 60}$$

$$= \frac{\pi r \cos 60}{9(\cos 40 - \cos 60)}$$

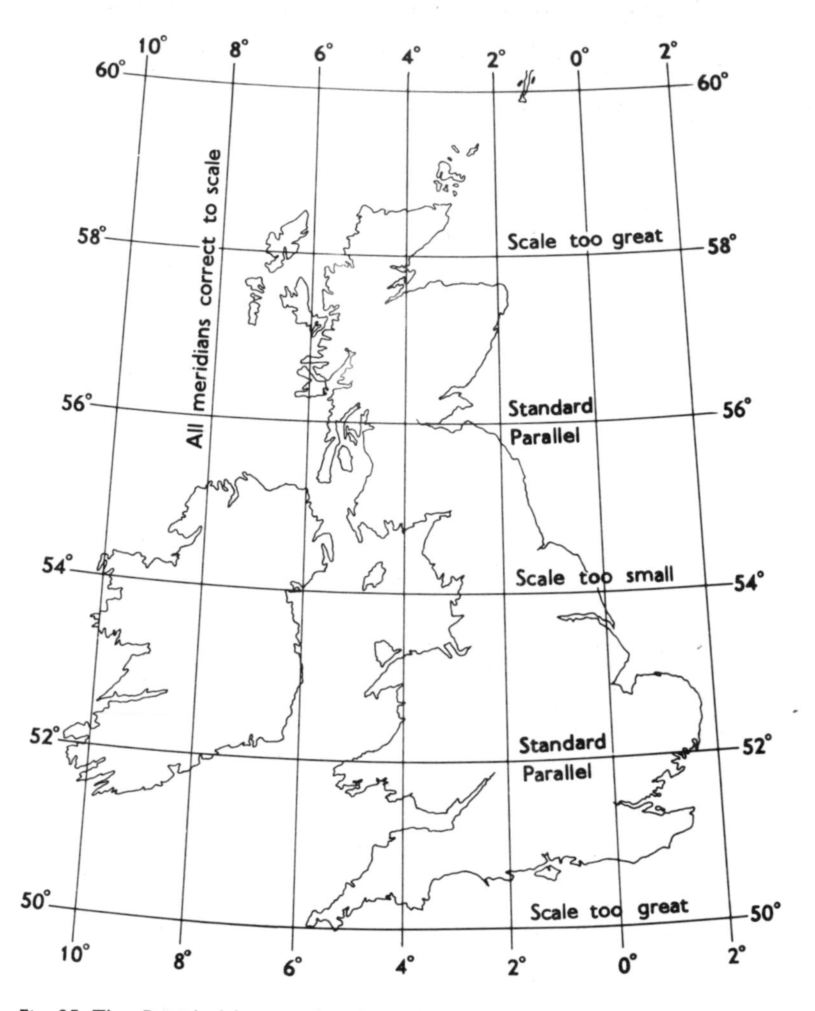

Fig. 25 The British Isles on the Conical with two standard parallels. (Scale 1 : 10 million or approximately 160 mile to 1 inch; radius of globe = 25 inches.)

A judicious choice of standard parallels is, however, necessary. A rule of thumb guide to their choice would fix them at about one-sixth of the total latitudinal extent of the area to be represented from both its northern and southern limits.

Uses

A severe limitation is placed on the usefulness of the Conical Projection with *One* Standard Parallel by the great and progressive exaggeration of its parallel scale, polewards and equatorwards of the standard parallel. It is, therefore, suitable only for countries with a small extent in latitude and preferably with a relatively great extent in longitude. The Conical with *Two* Standard Parallels suffers to some extent from the same defect but the spreading of the error of the parallel scale, by making some parallels too short and some too long, makes it suitable for representing areas of somewhat greater extent in latitude. Nevertheless, in most atlases, for example, Philips' *University Atlas*, edited by Fullard and Darby, it is used for countries rather than continents. (Fig. 25.) On pages 84 and 85 of this atlas, it will be noticed that it is also used for a map of the *U.S.S.R. in Europe* extending from 33°N. to 70°N., a range of approximately 35° in latitude.

9 Bonne's Projection

A very simple modification of the Conical with One Standard Parallel results in a projection called Bonne's whose equal-area property makes it capable of wide application. A suitable parallel is selected (see 'Choice of Standard Parallel' below and Fig. 27) and the construction proceeds through stages (*a*) and (*b*) on Fig. 26. So far, this is the same as the Conical with One Standard Parallel. However, in the Simple Conical, only *one* parallel, the standard parallel, is drawn correct to scale; it is subdivided correctly and the points of subdivision on it are joined to the centre of curvature of the standard parallel by straight lines which represent the meridians. In Bonne, *all* the parallels are correctly drawn and subdivided, i.e. if meridians are drawn at 10° intervals

the subdivision of each parallel is $\frac{2\pi r \cos \text{latitude}}{360} \times 10$ or $\frac{2\pi r \cos \text{latitude}}{36}$

After subdivision, the corresponding points on each parallel are joined by smooth curves to represent the meridians (Fig. 26(*c*) and (*d*)). They are composite curves rather than arcs of circles. Because all parallels are of correct length, the pole is represented by a point and not by an arc as it is on the Conical with One *and* the Conical with Two Standard Parallels.

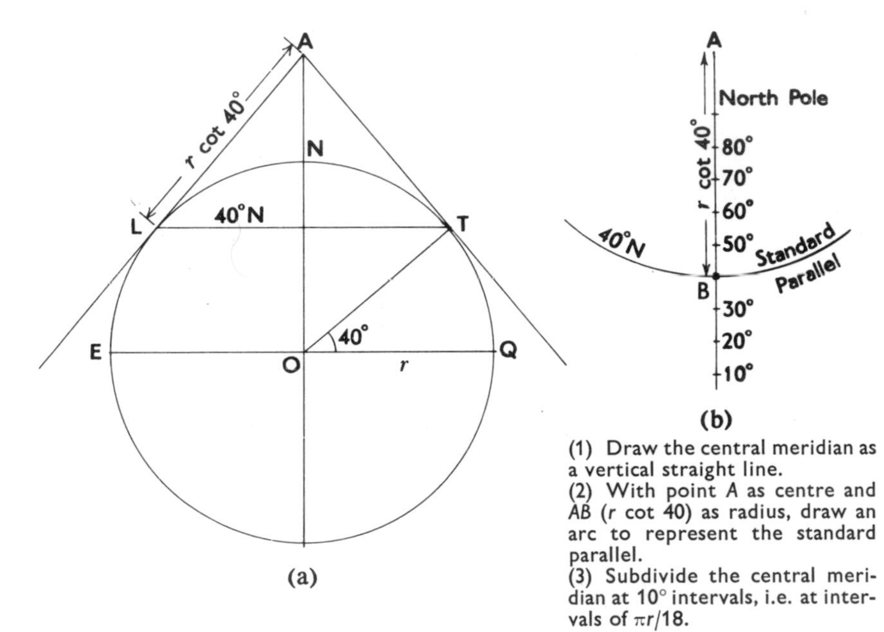

(a)

(b)

(1) Draw the central meridian as a vertical straight line.
(2) With point *A* as centre and *AB* (*r* cot 40) as radius, draw an arc to represent the standard parallel.
(3) Subdivide the central meridian at 10° intervals, i.e. at intervals of $\pi r/18$.

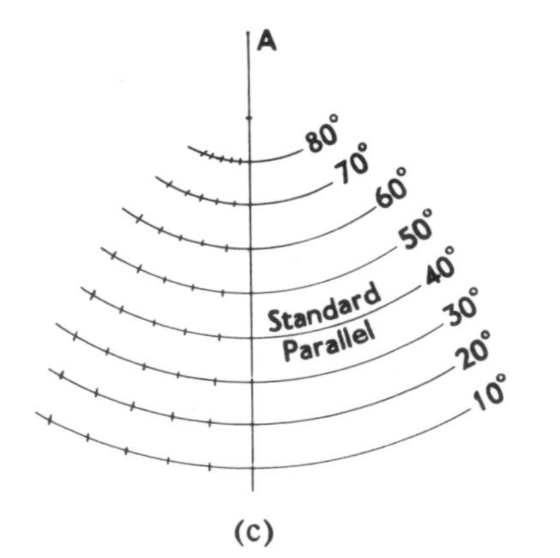

(c)

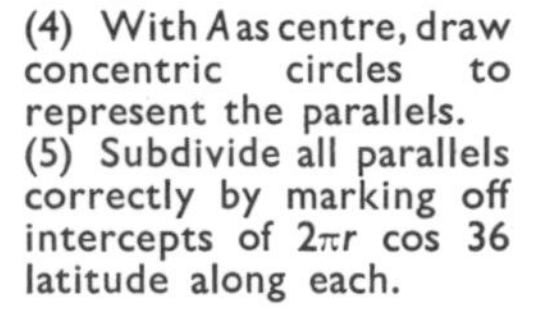

(4) With *A* as centre, draw concentric circles to represent the parallels.
(5) Subdivide all parallels correctly by marking off intercepts of $\frac{2\pi r \cos \text{latitude}}{36}$ along each.

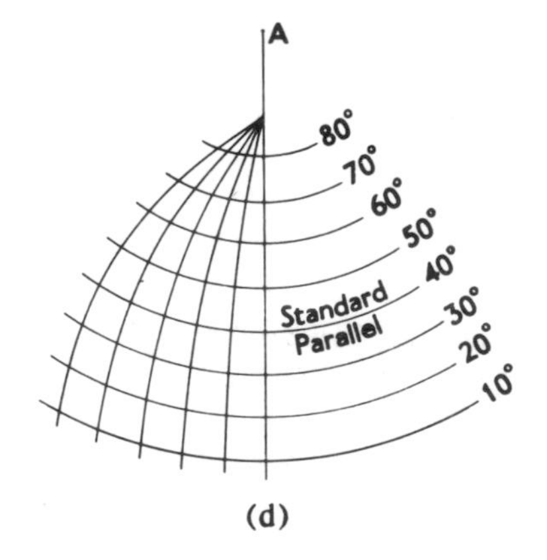

(d)

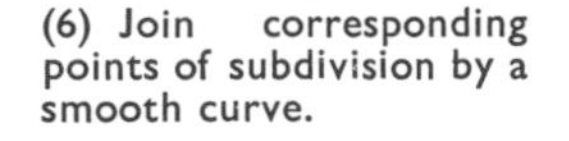

(6) Join corresponding points of subdivision by a smooth curve.

Fig. 26 Stages in the construction of Bonne's.

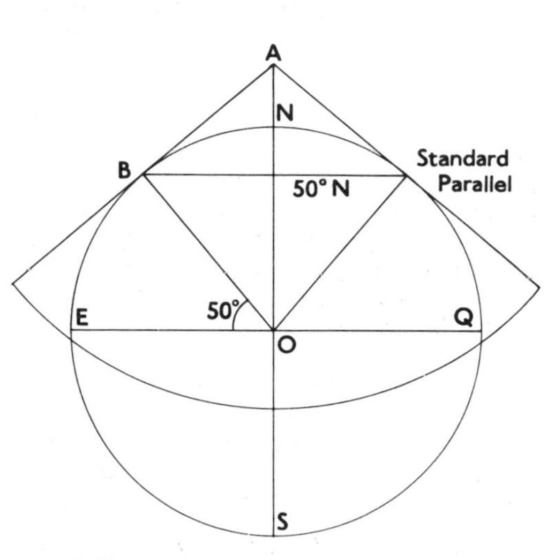

(*a*) The higher the angle of latitude of the standard parallel, the shorter is the radius, *AB*, of the standard parallel.

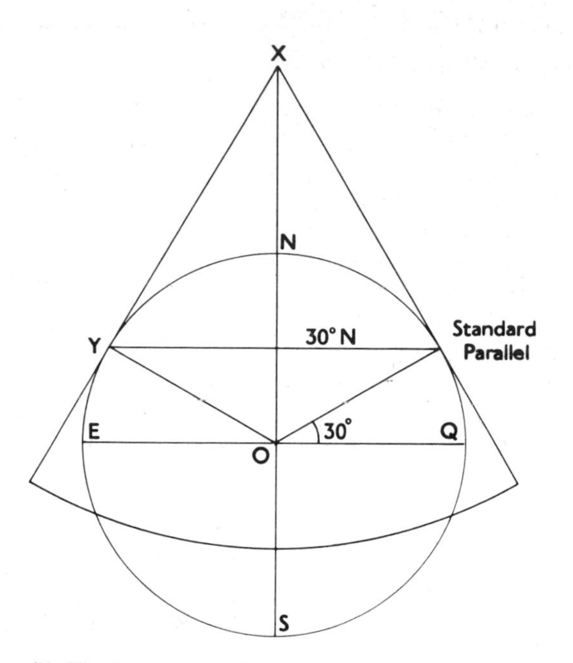

(*b*) The lower the angle of latitude of the standard parallel, the longer is the radius, *XY*, of the standard parallel.

Fig. 27 The effect of the choice of the standard parallel on the curvature of the parallels.

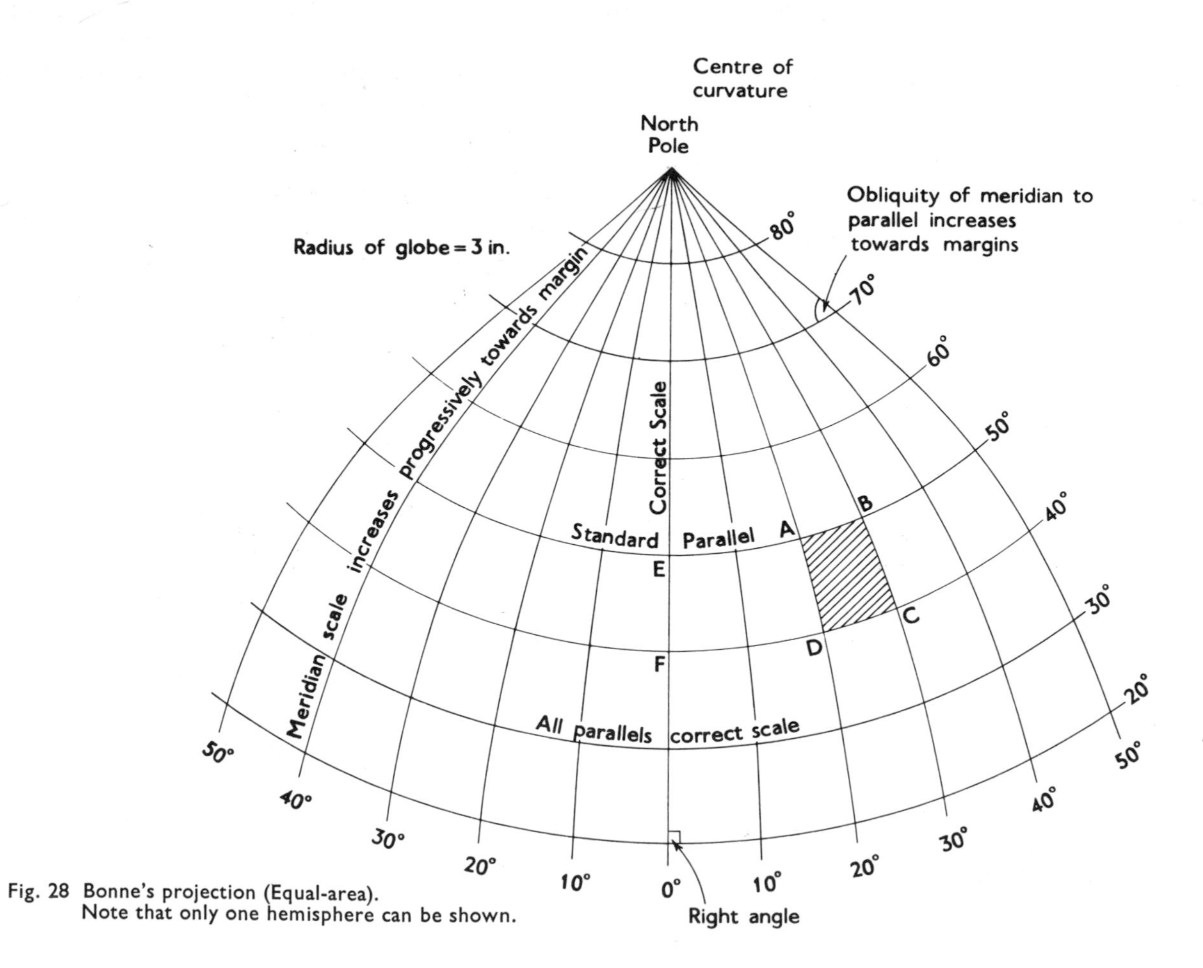

Fig. 28 Bonne's projection (Equal-area).
Note that only one hemisphere can be shown.

The Choice of Standard Parallel

The standard parallel chosen governs the degree of curvature of *all* the parallels. The nearer the equator it is, the greater is the radius of curvature and the 'flatter' the arc; the nearer the pole it is, the shorter is the radius of curvature and the 'rounder' the arc. (See Fig. 27.) The 'rounder' the arc of the parallels, the less is the obliquity between meridians and parallels in arctic latitudes and the greater it is in temperate and tropical latitudes. The 'flatter' the arc of the parallels, the less is the obliquity between the meridians and parallels in temperate latitudes and the greater it is in tropical and arctic latitudes. Wherever the obliquity is the greatest, the more rapid is the distortion of shape towards the margins of the map. Ideally, the standard parallel selected should pass through the centre of the area to be represented.

The following is a simple tabulated statement of the characteristics, properties, limitations and uses of the projection:

The meridians are composite curves meeting at the pole.

The parallels are concentric circles whose degree of curvature is dependent on the choice of the standard parallel.

Only the central meridian cuts the parallels at right angles. All others cut them obliquely, the obliquity being greatest towards the margins of the map.

Of the meridians, only the central meridian is drawn correct to scale. All others are progressively exaggerated in length towards the margins.

All parallels are correct to scale.

The projection is equal-area. This is readily seen to be so by a consideration of the area between any pair of parallels and any pair of meridians. In Fig. 28, for example, $ABCD$ is a trapezium in which AB and CD are the parallel sides.

The area of a trapezium is:

$$\frac{\text{the sum of the length of the parallel sides}}{2} \times \text{the perpendicular distance between the parallels—(in Fig. 28), } \frac{AB+DC}{2} \times EF.$$

These significant dimensions are all equal to their counterparts on the globe; so the trapezium is equal in area to the part of the globe it represents—and similarly, for any other part of the projection bounded by any pair of meridians and any pair of parallels. Hence the whole projection is equal in area to the portion of the globe it represents.

This equal-area property, however, can be achieved only at the expense of the shape of land-masses. Clearly, the obliquity of the meridians and parallels towards the margins of the map 'pulls' the land-masses out of shape. For example, on a map of Eurasia which has a longitudinal extent of more than 180°, Scandinavia on the one side and the Japanese archipelago with Sakhalin on the other are very much stretched from north to south because of the elongation of the meridians while they are both 'pulled out of vertical' by the acute-angled intersection between the meridians and parallels in those areas. The projection should, therefore, never be used to represent Eurasia but its equal-area property makes it very suitable and widely used in atlases for general-purpose and distribution maps of individual continents confined to one hemisphere—North America, Europe, Asia (sometimes including those parts of the United States of Indonesia beyond the equator), Australia or any part of these.

PART IV

TWO WORLD EQUAL-AREA PROJECTIONS

10 Sanson–Flamsteed's[1] (or Sinusoidal) Projection

Sanson–Flamsteed's projection (Fig. 29) may be regarded as a special case, in fact one of the limiting cases of Bonne's where the equator is the standard parallel. It *can* be used to represent the whole globe but the result is generally unsatisfactory as the shape of land-masses is badly distorted towards the margins of the projection for reasons given below.

On Sanson–Flamsteed, all parallels of latitude are drawn as parallel straight lines of correct length and their true distance apart. The central meridian, therefore, being of true length (πr) is half the length of the equator. It is the only meridian to cut the parallels at right angles.

All parallels are subdivided correctly and, through the corresponding points of subdivision of them meridians are drawn as composite curves—NOT arcs of circles—converging on the pole which is, of course, a point.

East and west of the selected central meridian which need not be the prime meridian, the meridians cut the parallels at progressively more acute angles. The meridians, consequently, become progressively longer towards the margins of the map.

Preservation of Shape

The great disadvantage of the projection—very bad peripheral distortion of shape of land-masses—springs from these two characteristic features of the meridians. On a world map drawn on a Sanson projection whose central meridian is the prime meridian, the western margins of North America and the eastern margins of Asia especially are so 'pulled out of upright' by the very great obliquity of the meridians to the parallels—especially in middle and high latitudes—as to cause very great marginal distortion of shape. The Japanese archipelago, the island of Sakhalin, the Korean and Kamchatka peninsulas and north-eastern Siberia—in fact the whole of the eastern seaboard of extra-tropical Asia are seen to be badly misshapen by being stretched and made narrow like a rubber band under tension. In the southern hemisphere the islands of New Zealand and the continent of Australia are similarly affected—as is extra-tropical South America—but to a smaller extent, because it is not so far west of Greenwich.

Representation of Area

The redeeming feature of Sanson–Flamsteed, however, is that like Bonne's and for the same reason, it is equal-area. As in Bonne's, this may be proved by considering the area of any part of the graticule bounded by a pair of parallels and a pair of meridians, for example, the trapezium $ABCD$ (Fig. 29).

The area of a trapezium is:

$$\frac{\text{the sum of the length of the parallel sides}}{2} \times \text{the perpendicular distance between the parallels}$$

—(in Fig. 29), $\frac{AB+DC}{2} \times EF$.

Because these significant dimensions are all equal to their counterparts on the globe, the trapezium $ABCD$ is equal in area to its counterpart on the globe. Since the same may be proved of any other part of the projection, the *whole* of the projection is equal in area to the globe it depicts.

Uses

When used to represent the whole globe, its usefulness is severely restricted by the serious distortion which it imposes on the shape of land-masses in temperate and arctic latitudes on the margins of the map—a distortion of shape which becomes progressively more serious with increasing distance from the central meridian as the meridians cut the parallels increasingly obliquely and thereby become increasingly exaggerated in length. In fact, for maps of the *whole* globe, it is now never used in atlases; but its equal-area property enables it to be used with good results for a map of a continent such as Africa which lies astride the equator and whose longitudinal extent (about 70°—from 20°W. to 50°E.) is not so great as to cause much marginal distortion of shape, provided of course that the central meridian chosen approximately bisects it longitudinally, i.e. at about 20°E. After all, the scale along all parallels is correct and so is that on the central meridian, while the longitudinal extent of the area to be mapped is so relatively small as to cause

[1] *Historical note:* John Flamsteed (1646–1719), a well-known British astronomer, became the first Astronomer Royal when Charles II founded the Royal Observatory at Greenwich in 1675. When seamen required a means of determining longitude accurately at sea, Flamsteed explained that it was impossible so long as the stars and the moon were so inaccurately charted; hence the establishment of the Royal Observatory. Regular observations begun by Flamsteed now form the basis of the Nautical Almanac which was first published in 1767. Sanson, however, a cartographer and geographer born in Picardie in 1600, was undoubtedly the originator of the projection which now bears the names of both men.

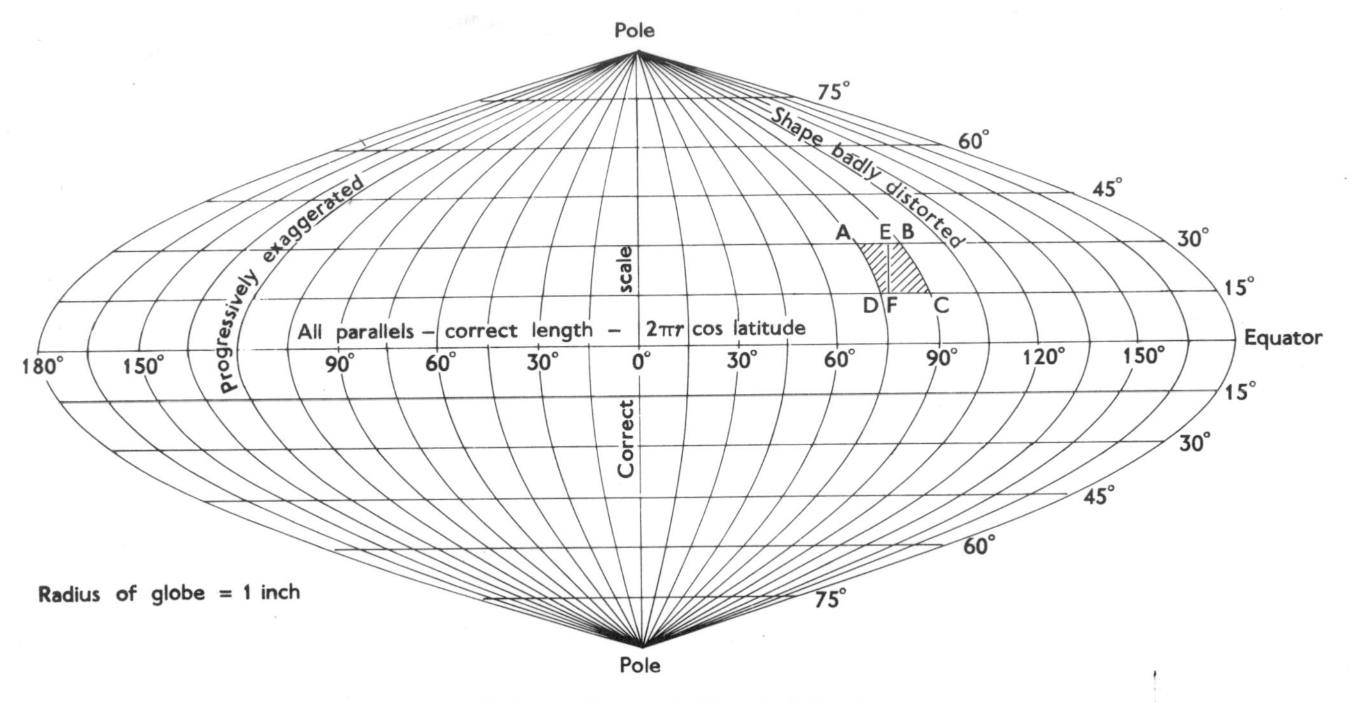

Fig. 29 Sanson–Flamsteed's (Sinusoidal) Equal-area.

only a slight departure from the right-angled intersection of all meridians and parallels, even at the eastern and western extremities; so, shape is reasonably well preserved.

Philips' *Modern School Atlas* uses it for separate physical maps of northern and southern Africa (pages 68–71) but maps of Africa as a whole are drawn on a smaller scale in this atlas on Lambert's Equivalent (i.e. Equal-area) Azimuthal projection, which is a popular one in many other atlases, for example, Bartholomew's *Comparative Atlas*, for maps of individual continents.

However, the representation of the shape of land-masses may be markedly improved on a map based on the Sanson–Flamsteed projection, even for the world as a whole, by the convenient device of recentring the projection over the continents and interrupting it over the oceans (Fig. 32). Since this device is often used to improve the representation of shape on maps using Mollweide's Homolographic projection, it is considered at the end of chapter 11.

11 Mollweide's Homolographic Projection

Mollweide's Homolographic or Equal-area projection, when used for the whole globe on one map, is an ellipse whose major axis, the equator, is twice the length of the minor axis, the central meridian.

The projection is deliberately made equal-area in the following way. (Fig. 30.)

(1) The central portion of the projection between 90°W. and 90°E. is a circle whose area is half that of the whole projection.

If the whole projection is to be made equal in area to a *sphere* (i.e. to $4\pi r^2$) the central circle, therefore, must be made equal in area to a *hemisphere* (i.e. to $2\pi r^2$); and, if the *radius* of the central circle is x, then it follows that:

$$\pi x^2 = 2\pi r^2$$

(Area of central circle) (Area of a hemisphere)

Therefore, $x^2 = 2r^2$ and $x = \sqrt{2r^2}$
$= 1{\cdot}414r$

(2) The length of the equator is made four times this length ($1{\cdot}414r$) by extending the diameter of the circle by $1{\cdot}414r$ at either end, so that the bounding ellipse can be drawn to enclose an area equal to the sphere it represents.

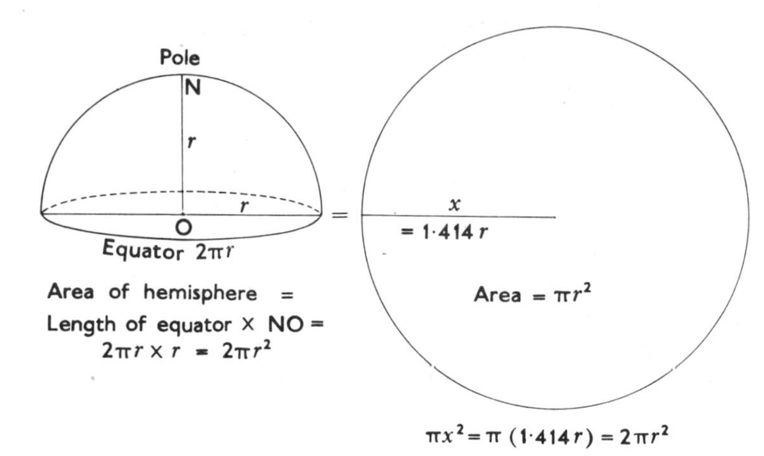

Fig. 30 The central circle on Mollweide is equal in area to a hemisphere.

(3) The equator is subdivided equally (but note, *not correctly*), and the meridians are drawn as ellipses through these points of subdivision and the poles. It follows that the areas between successive meridians (i.e. the gores) are all equal-area.

(4) Finally, the parallels are so spaced that the area between each (i.e. each zone) are all equal-area. This is mathematically the most difficult aspect of the projection.

Description (*Fig.* 31)

All parallels are straight lines. The length of the equator is $4 \times 1{\cdot}414r = 5{\cdot}656r$. This makes the equator shorter than that on the globe where it is $2\pi r$ or $6{\cdot}284r$. The pole is a point. Some of the remaining parallels are too long and others are too short, so that the error of the parallel scale is spread over the whole of the projection. Incidentally, one of the parallels must be the correct length; which one it is may be determined mathematically, but it is only of academic interest and need not concern us here.

All meridians except one are elliptical in shape. The central meridian is a straight line and, being *half* the length of the equator, is also shorter than it should be. Moreover, because the parallels become progressively closer together towards the poles, the scale along the central meridian diminishes progressively polewards.

Only the central meridian cuts the parallels at right angles. All others cut the parallels more acutely towards the margins of the map so that they become progressively longer. A point is therefore reached where one of the meridians on either side of the central meridian is the correct length. Because of the increasing obliquity of the meridians to the parallels, *especially in high latitudes* (temperate and arctic), the scales along each meridian increase towards the margins of the map and each meridian scale increases progressively polewards.

Representation of Shape

This increase of scale along the meridians together with the increasing obliquity between meridian and parallel towards the margins of the projection, results in one of the great disadvantages of the conventional Mollweide projection representing the whole of the globe (i.e. the *un*interrupted case). While shape is reasonably well preserved in the centre of the map, it becomes progressively worse towards the margins of the map as it does on Sanson–Flamsteed's. Landmasses, especially those in middle and high latitudes are 'pulled towards the poles' out of 'vertical' and they are, at the same time, elongated. If the central meridian is at Greenwich, the land areas of the globe which are most badly distorted in shape are Canada and, even more so, Alaska, on the west, northern Eurasia (and, especially, the eastern margins of Asia from Japan northwards to the Bering Straits), temperate South America, temperate Australia and, in particular, New Zealand.

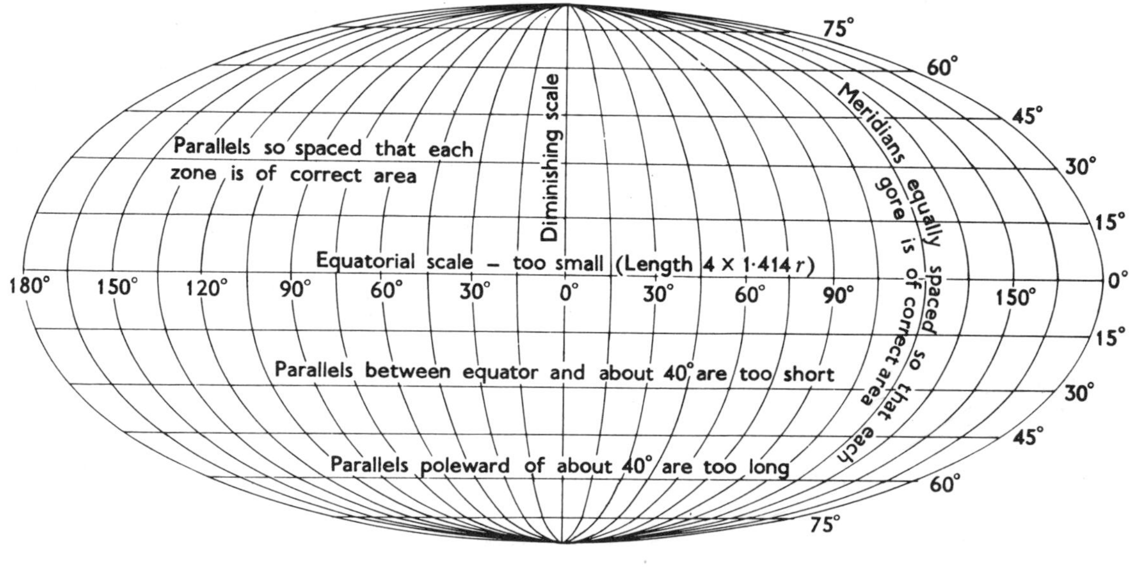

Fig. 31 Mollweide's Homolographic projection (Equal-area).

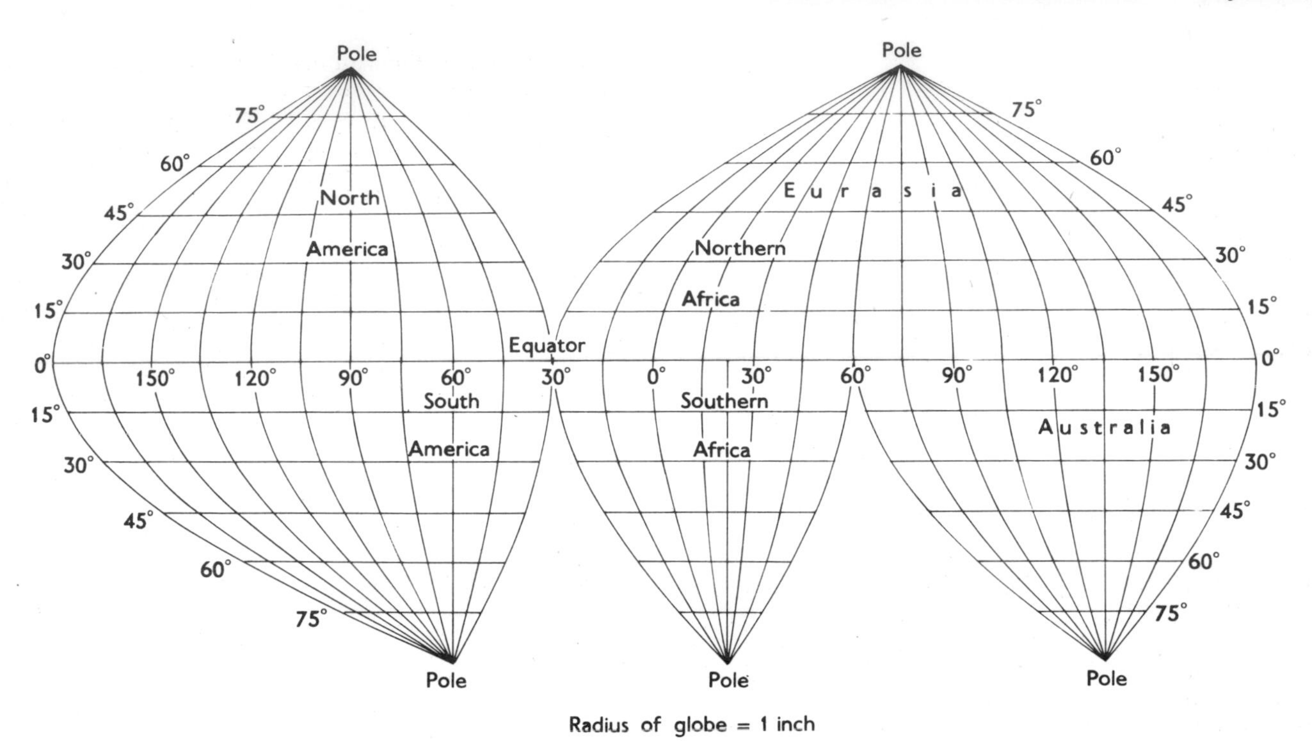

Fig. 32 The interrupted Sanson–Flamsteed.

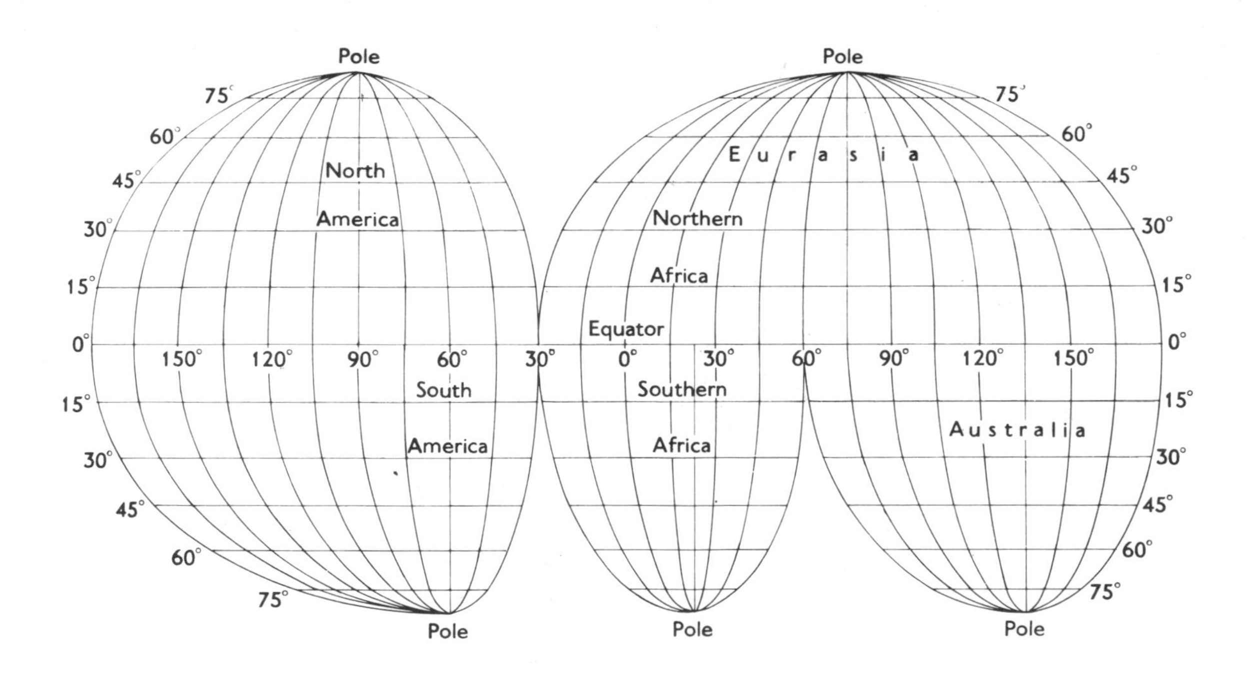

Fig. 33 The interrupted Mollweide.

However, the shape of land-masses on Mollweide is not so badly distorted as on Sanson–Flamsteed's simply because the equator on Mollweide is shorter than on Sanson–Flamsteed's, while some of the parallels are longer and others shorter than they should be. This has the effect of lessening the amount of obliquity between meridian and parallel with a consequent improvement in the shape of land-masses towards the margins.

Uses

As on Sanson–Flamsteed's, this marginal distortion of shape of land-masses severely restricts its usefulness in representing the whole globe; but, even for this purpose, it does give a better result than Sanson–Flamsteed does in providing a world map which, because of its equal-area property, is particularly valuable for showing distributions of various kinds. Whereas Sanson–Flamsteed's well represents land-masses such as Africa and South America which, besides having a relatively small longitudinal extent, lie astride the equator, Mollweide is less well suited to this purpose. On Sanson–Flamsteed's, all parallels and the central meridian have correct scales and, within 30° or so of the central meridian, the other meridians are not very much elongated. Mollweide does not possess this relative uniformity of scale along meridian and parallel.

Interruption and Recentring

The shape of land-masses on both Mollweide and Sanson–Flamsteed may be improved considerably by interrupting the projections over the oceans and by setting the most centrally placed meridians of each continental land-mass at right angles to the parallels. (Figs. 32 and 33.) This, of course, reduces the acuteness of the intersection of neighbouring meridians with the parallels so that the representation of shape is correspondingly improved. If Africa is in the centre of the map, the Atlantic Ocean is most conveniently interrupted at 20°W. and 30°W. in both hemispheres and the Indian Ocean at 60°E. or 80°E. The North American continent may be recentred around 90°W., Eurasia around 80°E.,[1] South America around 60°W., southern Africa around 20° E. and Australia around 130°E.

The parallels are spaced and subdivided exactly as they are in the conventional form of the projection so that the total length of the various parts of them is the same. Each of them is merely divided into a number of parts, *two* in the northern hemisphere, *three* in the southern to conform with the major land-masses. The Eurasian land-mass, however, still suffers from the marginal distortion of shape—Scandinavia, in particular and western Europe in general and the eastern seaboard of Asia, north of Japan, being most badly affected. The great longitudinal extent of the land-mass, from 10°W. eastwards to 180° makes this inevitable, since the interruption of the continental area is not practicable.

[1] Or around 20°E., if a better representation of shape is required for Africa rather than Eurasia.

MAP PROJECTIONS

H. S. Roblin

Erratum

Page 39: The two lower *diagrams* should be interchanged. The captions should remain in their present position.

PART V

ZENITHAL PROJECTIONS

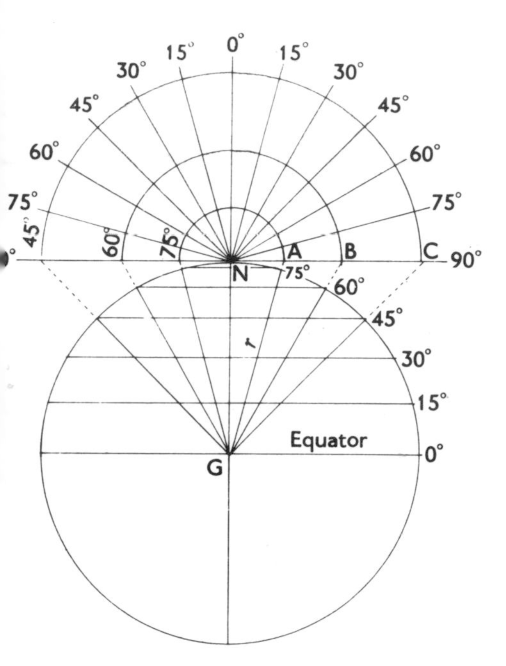

(*a*) Polar Zenithal Gnomonic.

(*a*) (1) Meridians and parallels are projected on to a plane tangent at the pole by a light at the centre of the earth *G*.
(2) Distance of any parallel from the pole, $N = r \cot$ latitude.
$NA = r \cot 75°$.
$NB = r \cot 60°$.
$NC = r \cot 45°$, etc.
(3) Parallels are drawn progressively further apart equatorwards.
(4) On it, any straight line is a great circle—the shortest distance between any two points. Direction from the pole is correct.

(*b*) (1) Meridians and parallels are projected on to a plane tangent at the pole by a light at the opposite pole, *S*.
(2) Distance of any parallel from the pole, $N =$

$$2r \times \tan \frac{\text{colatitude}}{2}$$

or

$$2r \times \tan \frac{90 - \text{latitude}}{2}.$$

(3) The spacing of the parallels increases progressively away from the pole, but not so rapidly as in the Gnomonic.
(4) The projection is orthomorphic. Direction from the pole is correct.

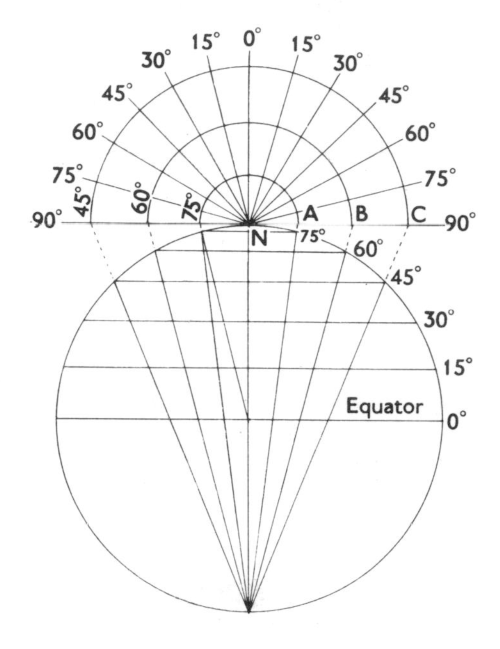

(*b*) Polar Zenithal Stereographic.

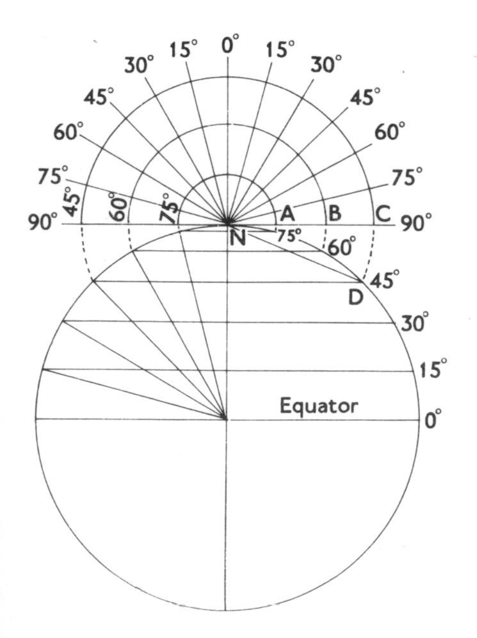

(*c*) Polar Zenithal Equidistant.

(*c*) (1) Meridians and parallels are not projected by an assumed source of light, i.e. it is a non-perspective projection.
(2) Distance of any parallel from the pole is correct, i.e. it is the arc distance on the globe of the parallel from the pole, i.e. *NC*, the distance of parallel 45° from the pole

$$= ND = \pi r \times \frac{\text{colatitude}}{180}.$$

(3) Parallels are equidistantly (and correctly) spaced.
(4) Both direction and distance from the pole are correct.

(*d*) (1) Meridians and parallels are not projected by an assumed source of light, i.e. it is a non-perspective projection.
(2) Distance of any parallel from the pole, $N =$

(i) $\sqrt{2r\,(r - r \sin \text{latitude})}$,

(ii) Alternatively, $2r \times \sin \frac{\text{colatitude}}{2}$

i.e. diameter of globe × sin colatitude = chord distance of parallel of latitude from the pole, e.g. $\frac{ND}{2}$.

(3) The spacing of the parallels decreases progressively away from the pole—but only slightly.
(4) The projection is equal-area. Direction from the pole is correct.

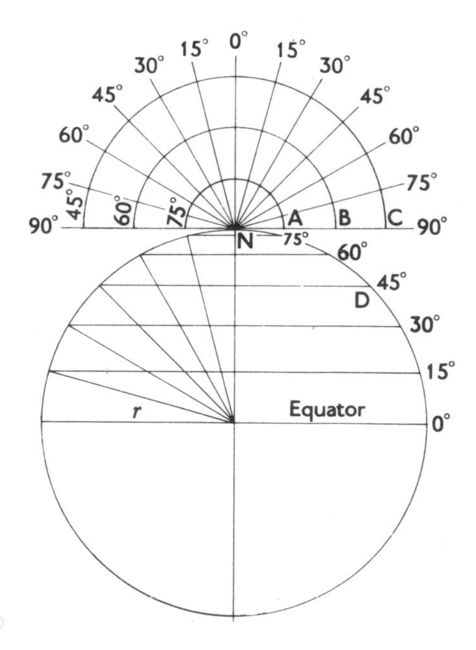

(*d*) Polar Zenithal Equal-area.

Fig. 34 Polar Zenithals.

In the examples of the cylindrical group of projections considered, a cylinder was assumed to be wrapped round the globe so as to touch it along the equator; in the conical group it was assumed that a cone was placed over the globe with its apex vertically above the pole so that it touched the globe along a *parallel of latitude* which was referred to as the *standard parallel.* In yet a third group, the zenithal, a plane (or sheet of paper) is assumed to touch the globe at a *single point.* It is further assumed that the globe is a transparent one on which the lines of latitude and longitude are marked so that a source of light inside the globe, at the edge of the globe or at infinity, may *project* them on to the plane surface.

Point of Contact

The plane may touch the globe at either *pole*, at the *equator* or at any *intermediate point* and the meridians and parallels will be projected on to the plane surface in a variety of ways to give rise to the Polar Zenithal (see page 39), the Equatorial Zenithal and the Oblique Zenithal respectively.

Source of Light

The form of the projected meridians and parallels and therefor the properties of the resulting projections will also vary according to the location of the source of light. If it is at the centre of the globe, the projection is called the Zenithal *Gnomonic*; if at a point on the surface of the globe but 180° away from the point of contact, the Zenithal *Stereographic* and, if at infinity, in which case the rays of light are deemed to be parallel, the Zenithal *Orthographic.*

By varying the position of the light source and of the point of contact of the plane surface with the globe, a multiplicity of projections may be derived; but, from a geographical point of view, many are valueless.

Besides the zenithal projections which are truly projected by an assumed source of light, i.e. the perspective types, there are a number of zenithals which are not projected and are, therefore, termed non-perspective; the most important of these are the Zenithal Equidistant (polar, equatorial and oblique cases) and the Zenithal Equal-area, often called Lambert's.

Of this bewildering assortment, we shall concern ourselves primarily with two major types, the Zenithal Gnomonic and the Zenithal Equidistant in their three major varieties, the polar, equatorial and oblique cases.

All zenithal projections have one major feature in common, that any straight line from the centre of the projection, that is from the point of contact of the plane with the globe, is a line of true direction or bearing.

12 The POLAR Zenithal Gnomonic Projection

This is one of the perspective zenithal projections. A source of light, assumed to be situated at the centre of the earth projects the meridians and parallels on to a plane surface which is tangent to the globe at either of the poles (in Fig. 35 at the North Pole).

As in all polar zenithal projections (Fig. 34(*a*) to (*d*), the pole is, therefore, the centre of the projection and parallels are represented as concentric circles with the pole as their common centre. It is obvious from Fig. 34(*a*) that because the source of light is at the centre of the earth, a ray of light from it through the equator would be parallel to a plane surface tangent at the pole and that the equator cannot be projected on to such a plane.

All meridians are shown as straight lines radiating from the pole, their true angular distance apart. Clearly, all meridians and parallels intersect at right angles because the meridians are the common radii of the concentric circles. It is evident from the portion of the projection shown (Figs. 34(*a*) and 35) that the parallels become progressively further apart with increasing distance from the pole and that, therefore, the scale along the meridians is not constant but increases progressively towards the equator:

Latitude	*Percentage exaggeration of scale along meridians*
75°	2·3%
60°	10·0%
45°	27·4%
30°	165·4%

(See 'Calculation' at end of chapter.)

Clearly, within 30° of the pole, the exaggeration of 10% along the meridian is relatively small, but equatorwards of this the amount of stretching of the meridian increases very rapidly.

Because of the increasing length of their radii towards the equator, the parallel scale is exaggerated increasingly equatorwards. The rate of exaggeration is much greater than along the meridians, as the following table shows:

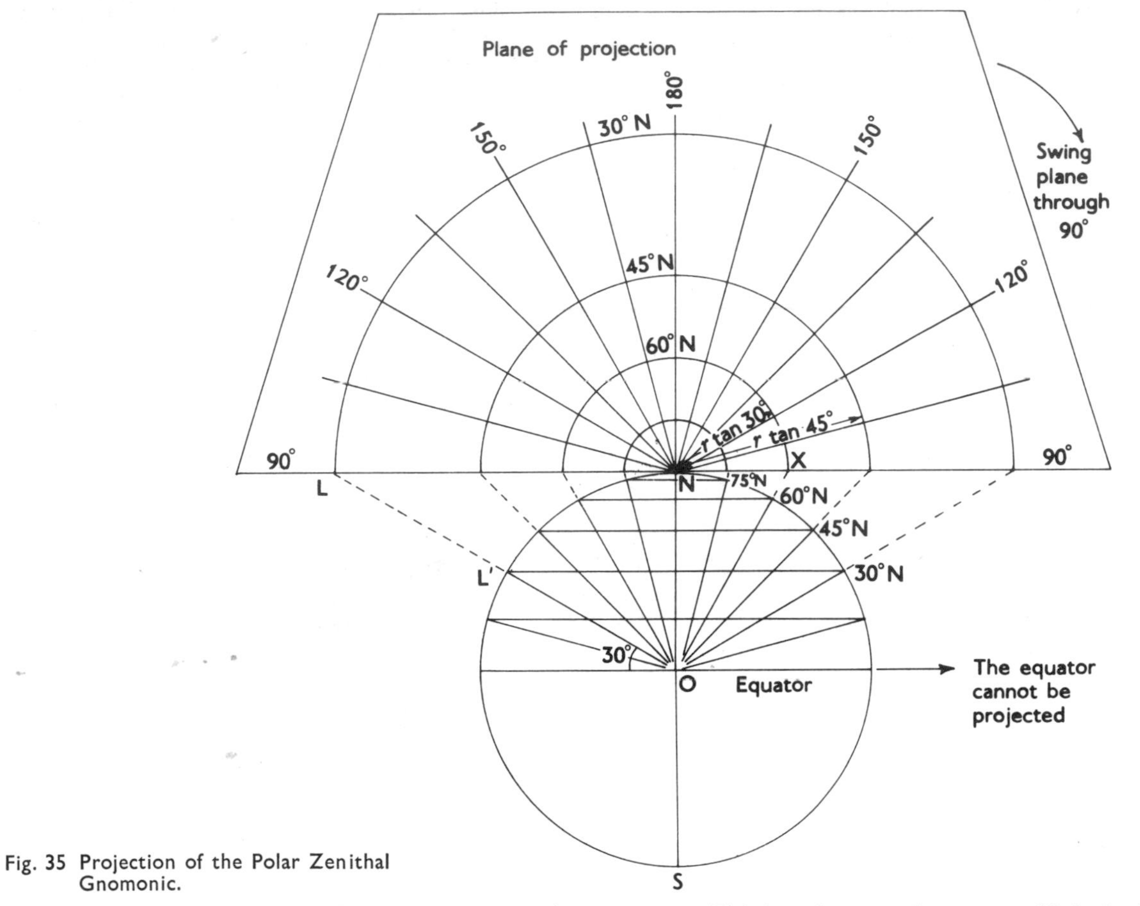

Fig. 35 Projection of the Polar Zenithal Gnomonic.

Latitude	*Percentage exaggeration of scale along the parallels*
75°	3·5%
60°	15·5%
45°	41·5%
30°	200·0%

The projection is, obviously, not orthomorphic because although meridians and parallels intersect at right angles, at no point on the projection is the scale along the meridian and parallel the *same*. As has been illustrated above, at any point one cares to select, the meridian scale is exaggerated and so is the parallel scale but by a greater amount.

For this reason too, the projection cannot be equal-area; but from the above tables of percentage exaggeration of the scales along both meridian and parallel, it is clear that within 30° of the pole, the amount of stretching of the meridians (10%) and the parallels (15%) is so relatively small that shape and area are reasonably well represented in arctic areas.

The most valuable property of all three cases of the zenithal gnomonic—the polar as well as the equatorial and the oblique—is that all great circles appear as straight lines.[1]

[1] In relation to meridians, this is easy to see because the meridians clearly lie on great circles and they are *seen* to be represented by straight lines (see Fig. 35); but imagine yourself situated at the centre of a large globe, i.e. at the point from which the meridians and parallels are projected, and try to visualize how an 'oblique' great circle would be projected by a light from your position on to a plane surface which is tangent at the pole.

This is so because the source of light is situated *at the centre of the earth* while, at the same time, by definition, the planes of great circles must pass *through the centre of the earth.* To find the shortest route between one point and another, all one has to do is to join the two points by a straight line. This property, in combination with the accuracy of direction from the centre of the projection makes the polar zenithal gnomonic a most valuable projection for navigation; in addition, the reasonable accuracy of meridian and parallel scale and, therefore, of shape and area within 30° of the pole makes it suitable for general-purpose maps of arctic areas.

Calculation

(1) *To find distance of parallels, e.g.* 30°*N. or* 30°*S. from the pole* (*Fig.* 35)

In $\triangle\ ONL$, $\angle ONL = 90°$, $NO = r$, and $\angle NOL$ = angle of colatitude = 60°

$\frac{NL}{NO}$ = tan colatitude (60°).

Therefore, $NL = NO$ tan colatitude 60° = r tan colatitud 60°.

Similarly for any other parallel of latitude.

(2) *To find percentage exaggeration of meridian scale at any latitude, e.g.* 30°*N. or* 30°*S.*:

Assume radius of globe to be 1 inch.

(*a*) Distance of parallel 30°N. from pole ON GLOBE $= 60°$ of arc

$$= \frac{2\pi r}{360} \times 60 = \frac{2\pi r}{6} = \frac{\pi r}{3} = \frac{3{\cdot}142}{3} = 1{\cdot}047 \text{ inches.}$$

(*b*) Distance of parallel 30°N. from pole ON PROJECTION $= r \tan$ colatitude $= r \tan 60° = 1{\cdot}7321$ inches.

(*c*) Therefore, percentage exaggeration along meridian at 30°N.

$$= \frac{\text{Distance of parallel 30°N. from pole ON PROJECTION}}{\text{Distance of parallel 30°N. from pole ON GLOBE}} \times 100$$

$$= \frac{1{\cdot}7321}{1{\cdot}047} \times 100 = \frac{173{\cdot}21}{1{\cdot}047} = 165{\cdot}4\%$$

i.e. 65·4% exaggerated.

No.	Log.
173·21	2·2385
1·047	0·0199
	2·2186
Antilog.	165·4

Similarly for any other parallel.

(3) *To find percentage exaggeration of parallel scale at any latitude, e.g.* 30°*N.* or 30°*S.:*

(*a*) Length of parallel 30°N. ON PROJECTION $=$ $2\pi \times$ radius of parallel 30° ON PROJECTION.

Radius of parallel 30° ON PROJECTION $= r \tan$ colatitude $= r \tan 60°$.

Therefore, length of parallel 30° ON PROJECTION $=$ $2\pi \times r \tan 60°$.

(*b*) Length of parallel 30° ON GLOBE $= 2\pi r \cos 30$.

(*c*) Therefore, percentage exaggeration of parallel 30° $=$

$$\frac{\text{Length of parallel 30° ON PROJECTION}}{\text{Length of parallel 30° ON GLOBE}} \times 100 = \frac{2\pi r \tan \text{colatitude}}{2\pi r \cos \text{latitude}} \times 100 = \frac{2\pi r \tan 60}{2\pi r \cos 30} = \frac{\tan 60}{\cos 30} \times 100 =$$

$$\frac{1{\cdot}7321}{0{\cdot}866} \times 100 = \frac{173{\cdot}21}{0{\cdot}866} = 200\%$$

i.e. 100% exaggerated (or doubled).

No.	Log.
173·21	2·2385
0·866	$\bar{1}$·9375
	2·3010
Antilog.	200·0

Similarly for any other parallel.

13 The EQUATORIAL Zenithal Gnomonic Projection

In the Equatorial Zenithal Gnomonic, a light situated at the centre of the globe projects the meridians and parallels of the globe on to a plane touching the globe (i.e. tangent to it) at the equator. (Fig. 36.)

As in the *Polar* Zenithal Gnomonic, the greatest asset of the *Equatorial* Zenithal Gnomonic is that all great circles are projected as straight lines. So the equator and the meridians, being great circles, are represented as straight lines. The meridians are projected at right angles to the equator and are consequently parallel to one another. (See Figs. 36 and 37.)

On Fig. 36, *P* is a point on the equator which is the point of contact of the plane of projection with the globe. The diagram attempts to show in perspective that as one proceeds eastwards (or westwards) from *P:*

(*a*) the parallel, straight-line meridians become further apart and that

(*b*) points along the parallels at increasing distance from the central meridian become progressively further away from the equator. The parallels, for example, *L'A'T'* are therefore curved lines—but *not* arcs of circles, and each parallel becomes more curved polewards. (See Fig. 36.)

Those interested in the mathematical calculations necessary for constructing the projection, i.e. for determining the positions of the intersections of the meridians and parallels should refer to 'Calculations' at the end of the chapter.

Whereas in the polar case the *equator* cannot be projected by a light from the centre, it is evident that the *pole* cannot be projected in the equatorial case.

Clearly, except at the equator, the parallels do not cut the meridians at right angles, but as one proceeds polewards along each meridian the angle of intersection of meridian and parallel becomes progressively more acute.

The scale along the equator increases progressively eastwards and westwards of the centre of the projection *as the tangent of the angle of longitude*. Similarly, the scale along the central meridian increases progressively polewards *as the tangent of the angle of latitude*. The scale along the other meridians also increases polewards and this becomes progressively more pronounced towards the margins of the map.

It is obvious that the projection is neither equal-area nor orthomorphic because of (*a*) the progressive exaggeration of both meridian and parallel scale away from the centre of the map and (*b*) the increasing obliquity of the meridians to the parallels polewards.

The projection represents land-masses most accurately both in shape and in area nearer its centre and, because of this, it is most suitable for representing land-masses such as Africa which lie astride the equator and do not extend much

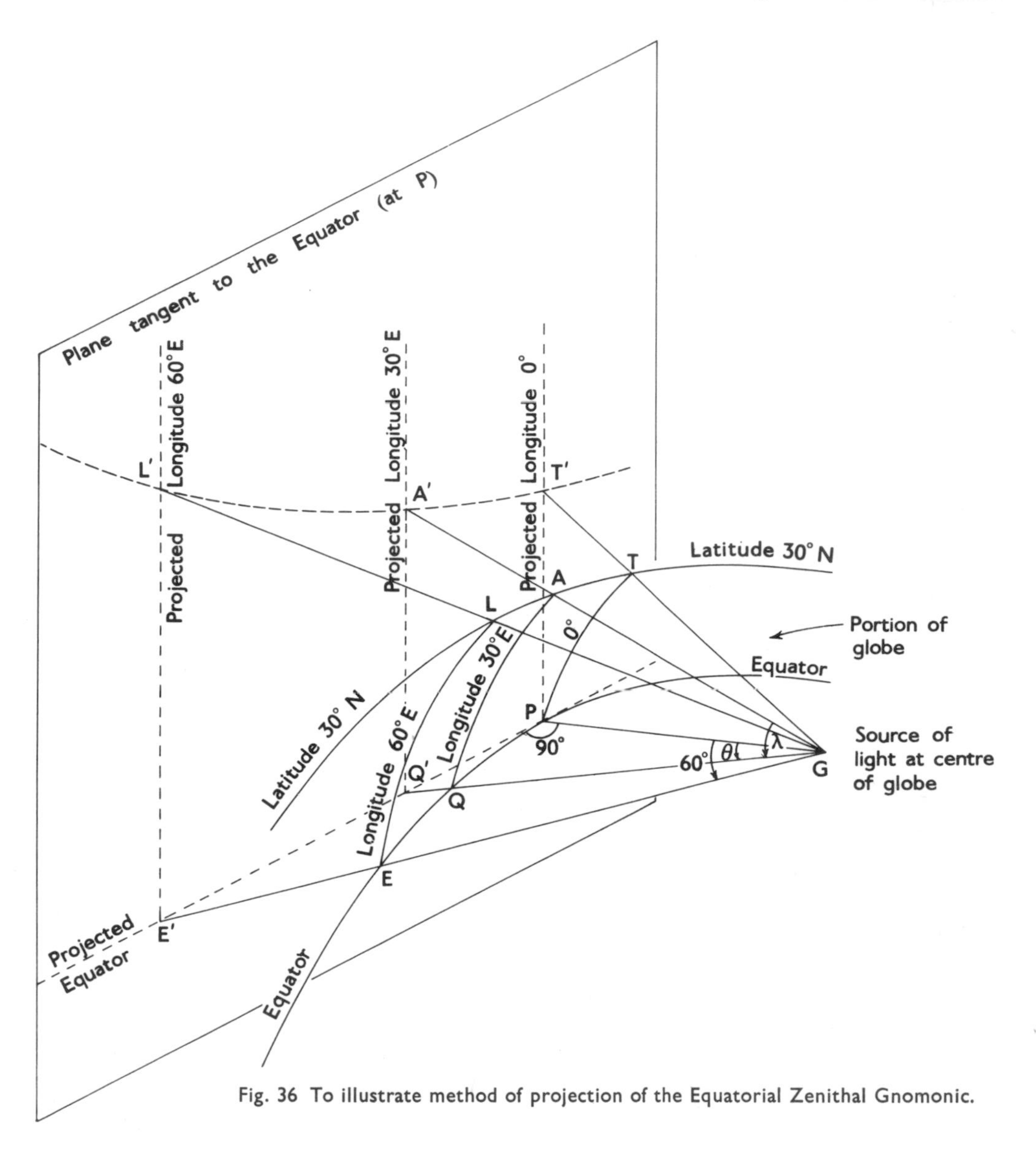

Fig. 36 To illustrate method of projection of the Equatorial Zenithal Gnomonic.

more than 30° from the centre of the map in any direction. The projection used for such a map would have the following advantages:

(1) Shape *and* area would be reasonably well represented.
(2) Direction or bearing from the centre of the map (i.e. the point of contact of the plane of projection with the globe) would be represented correctly.
(3) Any straight line drawn on the map would be a great circle.

Calculation

(1) *To find distance of meridians (e.g.* 30°*E.) from the central meridian (see Fig.* 36)

In $\triangle Q'PG$, G is the centre of the earth,
P is the point of contact of the plane of projection with the globe,
PG is the radius of the globe,
Q is a point on the equator, 30°E. of P, and
$\angle Q'PG$ is 90°.

$$\frac{Q'P}{PG} = \tan\theta;\ Q'P = PG \times \tan\theta = r\tan\theta.$$

(2) *To find the position of the point of intersection of any parallel with any meridian* (e.g. point A' at the intersection of parallel of latitude 30°N. with meridian 30°E.)

a) In $\triangle Q'PG$, $\dfrac{Q'G}{PG} = \text{secant}\ \theta$; $Q'G = r$ secant θ.

(*b*) In $\triangle A'Q'G$, angle $A'Q'G = 90°$; angle $A'GQ' = \lambda = 30°$ and $Q'G = r$ secant θ.

Therefore, $\dfrac{A'Q'}{Q'G} = \tan\lambda$; $A'Q' = Q'G \tan\lambda$
$= r$ secant $\theta \times \tan\lambda$

where θ is the number of degrees east (or west) of the point of contact, and λ is the angle of latitude.

Similarly for any other point of known latitude and longitude.

(3) (*a*) *To find the percentage exaggeration of the scale along the equator at any point* (e.g. at Q', 30° away from the point of contact)

$$\text{Percentage exaggeration} = \frac{Q'P}{\underset{\sim}{Q}P} \times 100$$

$Q'P = r \tan 30°$;

$$QP = 30° \text{ of arc } = \frac{2\pi r}{360} \times 30 = \frac{2\pi r}{12} = \frac{\pi r}{6}$$

If radius of globe = 1 inch, then:

$$\frac{Q'P}{QP} \times 100 = \frac{r \tan 30}{\pi r/6} \times 100 = \frac{\tan 30}{3{\cdot}142/6} \times 100$$
$$= 110{\cdot}3$$

i.e. about 10% exaggerated.

No.	Log.
57·74	1·7615
0·5236	$\bar{1}$·7190
	2·0425
Antilog.	110·3

(*b*) *To find percentage exaggeration at E', 60° away from point of contact*

$E'P = r \tan 60$

$$EP = 60° \text{ of arc} = \frac{2\pi r}{360} \times 60 = \frac{2\pi r}{6} = \frac{\pi r}{3}$$

If radius of globe = 1 inch, then:

$$\frac{E'P}{EP} \times 100 = \frac{r \tan 60}{\pi r/3} \times 100 = \frac{\tan 60}{\pi/3} \times 100$$
$$\frac{173{\cdot}21}{1{\cdot}0473} = 165{\cdot}4\%$$

i.e. 65·4% exaggerated.

(4) Similarly, the scale along the central meridian at T', latitude 30°N. is exaggerated by the same amount as in (3)(*a*), i.e. about 10%.

(5) *To find the exaggeration of scale along a meridian* (60°E. *of the point of contact*) *at latitude* 30°*N*. (i.e. at L' in diagram)

$L'E' = r \text{ secant } 60 \times \tan 30$

$$LE = 30° \text{ of arc} = \frac{2\pi r \times 30}{360} = \frac{2\pi r}{12} = \frac{\pi r}{6}$$

If radius of globe = 1 inch,

$$\frac{L'E'}{LE} = \frac{r \sec 60 \times \tan 30}{\pi r/6} \times 100 = \frac{\sec 60 \times \tan 30}{\pi/6} \times 100$$
$$= \frac{2{\cdot}0 \times 0{\cdot}5774}{0{\cdot}5236} \times 100 = 175{\cdot}2\%$$

i.e. 75·2% exaggerated.

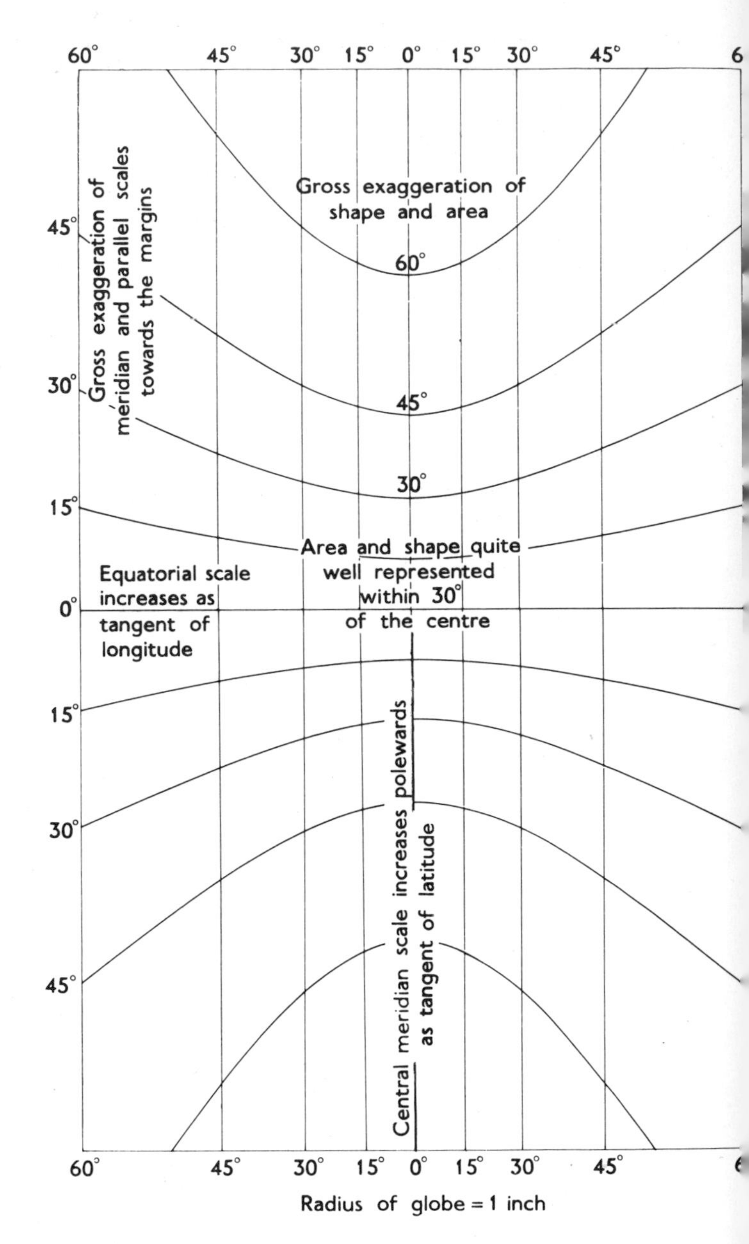

Fig. 37 The Equatorial Zenithal Gnomonic.

14 Zenithal Equidistant Projections

In all projections of this group, the most valuable properties are that *both direction and distance from the centre of the map are correct.*

To achieve this object, limitations are placed on their usefulness in other respects.

(1) only one-half of the globe can be shown;
(2) there is great distortion of shape along the margins of the projections; and
(3) area is progressively exaggerated from the centre outwards.

The Polar Zenithal Equidistant (*Figs.* 34(c) *and* 38)

Meridians are straight lines radiating from the pole and drawn their true angular distance apart.

Parallels are concentric circles whose common centre is the pole and drawn their correct distance apart, i.e. at intervals of $\frac{\pi r}{18}$ if drawn at 10° intervals.

This means that the scale along the meridians is correct.

The effect of the curvature of the earth is eliminated because distances along the meridians are represented as straight lines on a plane surface. Consequently, there is *an exaggeration in the length of the parallels* which increases progressively away from the pole. The amount of the exaggeration of the scale along each parallel can be calculated quite easily as follows:

$$\frac{\text{Length of parallel ON PROJECTION}}{\text{Length of parallel ON GLOBE}}$$

(See end of chapter.)

Notice that equatorwards of about 40° and 45° latitude, the exaggeration along the parallels increases very rapidly because it is *beyond* these latitudes that, from a viewpoint at the pole, the curvature of the earth appears most pronounced. The exaggeration of scale along the equator would be:

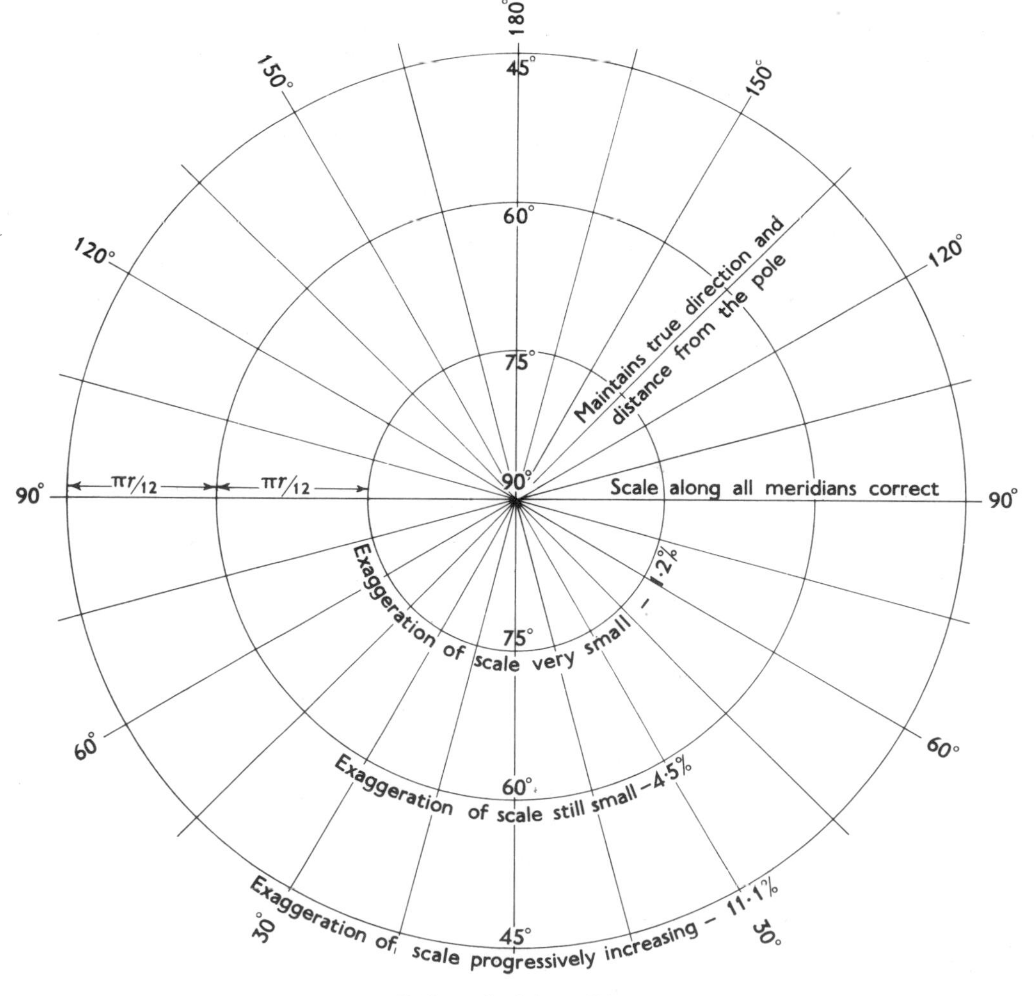

Fig. 38 The Polar Zenithal Equidistant.

$$\frac{\text{Length of equator ON PROJECTION}}{\text{Length of equator ON GLOBE}}=\frac{2\pi\times\pi r/2}{2\pi r}$$

$=\frac{\pi}{2}=\frac{3\cdot142}{2}=1\cdot57$ times

i.e. 57% exaggeration.

The projection is obviously not orthomorphic because, although the meridians intersect the parallels at right angles, the scale along all the meridians is correct while the scale along the parallels is progressively exaggerated away from the pole. Consequently, at no point on the projection are the scales along both meridian and parallel the *same*.

Neither is the projection equal-area because at no point are both meridian and parallel scales correct, nor are they compensatory. However, both shape and area are reasonably well preserved within about 30° of the pole and the projection can be used for general-purpose maps of arctic areas; but the *most useful properties* of the projection are the preservation of both distance and direction *from the centre of the map only*. This makes it suitable for polar exploration and, to some extent, for polar navigation.

Calculation

To calculate the exaggeration of scale along any parallel

(*a*) Length of any parallel ON PROJECTION $=2\pi\times$ radius of circle representing parallel ON PROJECTION (i.e. its distance from the pole).

(*b*) Length of any parallel ON GLOBE $=2\pi r$ cos latitude.

Exaggeration of scale $=$

$$\frac{\text{Length of parallel ON PROJECTION}}{\text{Length of parallel ON GLOBE}}$$

Exaggeration

(i) At 75°N.

$$\frac{2\pi\times\pi r/12}{2\pi r\cos 75}=\frac{\pi}{12\times\cos 75}=\frac{3\cdot142}{3\cdot1056}=1\cdot012,\text{ i.e. }1\cdot2\%$$

(ii) At 60°N.

$$\frac{2\pi\times\pi r/6}{2\pi r\cos 60}=\frac{\pi}{6\times\cos 60}=\frac{3\cdot142}{3}=1\cdot045,\text{ i.e. }4\cdot5\%$$

(i) and (ii): Very small

(iii) At 45°N.

$$\frac{2\pi\times r/4}{2\pi r\cos 45}=\frac{\pi}{4\times\cos 45}=\frac{3\cdot142}{1\cdot828}=1\cdot111,\text{ i.e. }11\cdot1\%$$

(iv) At 30°N.

$$\frac{2\pi\times r/3}{2\pi r\cos 30}=\frac{\pi}{3\times\cos 30}=\frac{3\cdot142}{2\cdot598}=1\cdot209,\text{ i.e. }20\cdot9\%$$

(v) At 15°N.

$$\frac{2\pi\times 5r/12}{2\pi r\cos 15}=\frac{5\pi}{12\times\cos 15}=\frac{3\cdot142}{2\cdot3184}=1\cdot355,\text{ i.e. }35\cdot5\%$$

(iv) and (v): Great

The Equatorial Zenithal Equidistant (*Fig.* 39)

As in the polar case, the projection can cover only half the globe but in this case the centre of the projection is any selected point *on the equator*.

The meridians are all arcs of circles whose centres are points either on the *equator* or the *equator produced*; they all converge on the poles. The parallels are also arcs of circles whose centres are points on the *central meridian produced*; but because each parallel has its own centre, they are not concentric.

The central meridian and the equator are straight lines of correct length intersecting at right angles.

All other meridians and parallels are too long, but each one is subdivided *equally*—hence the name 'equidistant'; all parallels are equidistant from each other and so are all meridians. Away from the *central* meridian, all others are progressively exaggerated towards the margins. The peripheral meridians on either side are 1·57 times too long, i.e. the ratio of the central meridian (which is correct) to the peripheral meridians is as diameter to semicircle, i.e. $2r:\pi r$ or $1:1\cdot57$.

The equator is the correct length and, being represented by a point, so is the pole. Intermediate parallels are *nearly* correct to scale.

Only the *central* meridian cuts the parallels at right angles. The only parallel to cut the meridians at right angles is the equator. All other meridians cut the parallels progressively more acutely towards the margins of the projection, but the intersection of meridians and parallels is never very oblique because of the curvature of the parallels towards the poles.

The projection is not orthomorphic because at no point are the scales along both meridian and parallel the same *except* at the centre of the map where they are both correct, nor do the meridians and parallels intersect at right angles *except* along the central meridian and the equator.

The projection is not equal-area because while the scales along the parallels are nearly correct, those along the meridians are progressively exaggerated towards the margins.

Uses

As in the *Polar* Zenithal Equidistant, the most useful property is that both distance and direction (or bearing) are preserved *from the centre* of the map. It can be used, therefore, for aerial navigation from any *one airfield located on the equator*, for example, Entebbe or Singapore. All straight lines drawn from the centre of the map are both great circles and lines of true bearing; but these desirable properties are achieved by the very great sacrifice of scale, area and shape towards the margins of the map. The uses of the equatorial case for air navigation are rather limited because large international airports on or very near the equator are few in number.

The Oblique Zenithal Equidistant

This is the most useful of the Zenithal Equidistant projections because it has a much wider application than the polar or equatorial cases. Its centre can be any point on the earth's surface intermediate between the pole and the equator.

As in the other cases, only half the globe can be shown. The whole projection is bounded by a circle whose radius is $\pi r/2$, where r is the radius of the globe. The vertical diameter is composed of portions of two meridians which are 180° apart and the parts of both meridians represented are drawn to scale. If, for example, the projection is centred on London, 51° 30′N., the vertical diameter will consist mainly of the Greenwich Meridian from the pole *almost* to 40°S. and the remainder of it will represent that part of meridian 180° from the pole to just *beyond* 40°N.

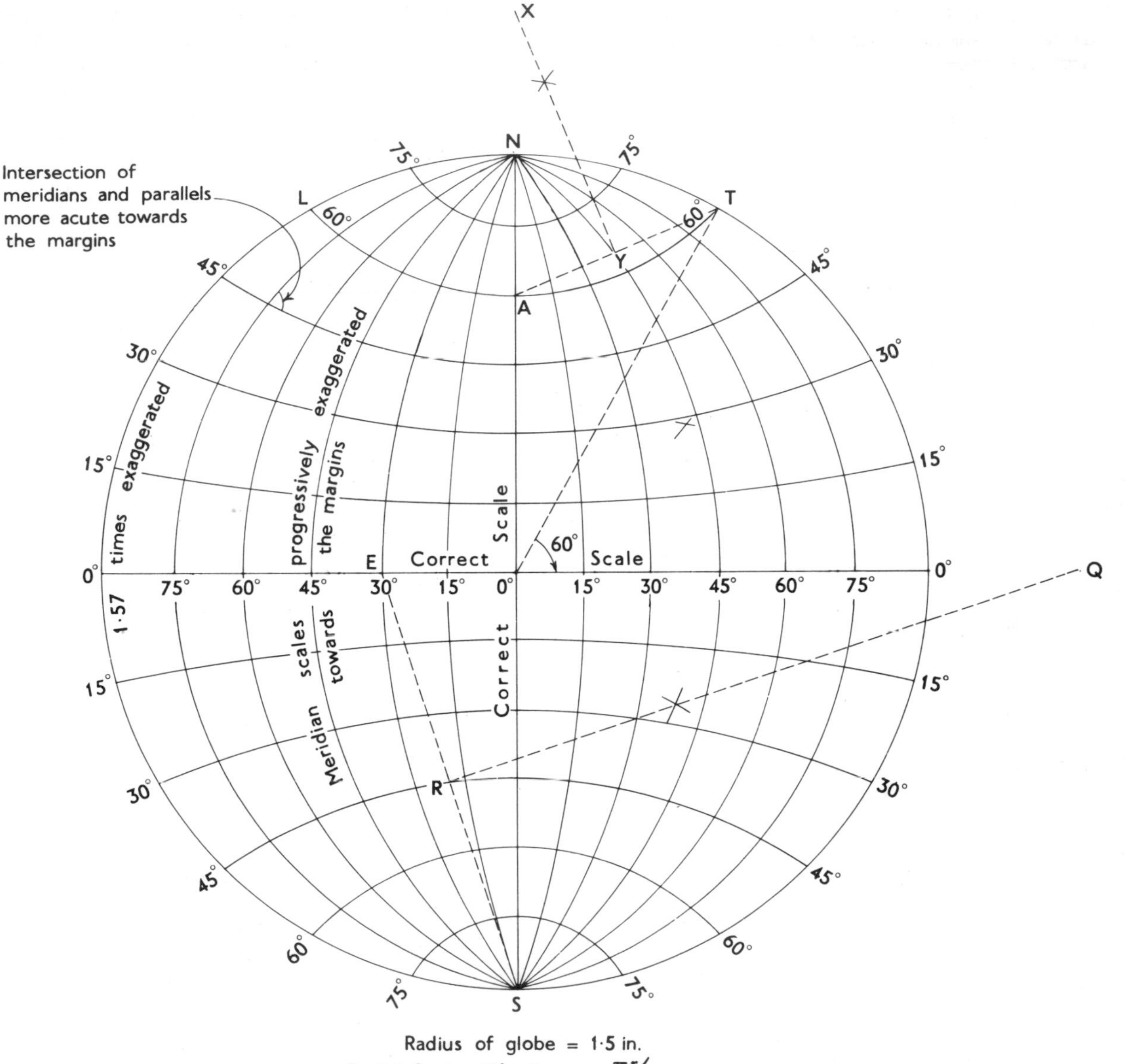

Fig. 39 The Equatorial Zenithal Equidistant.

The scale along the vertical diameter or central meridians is correct; both are subdivided equally and, therefore correctly. Each subdivision will be $\pi r/9$, if parallels are drawn at 20° intervals. Some of the parallels are shown complete, 80°N., 60°N. and 40°N. on the map centred on London, but they are ellipses rather than circles; each one is elongated east to west and they are not parallel to one another. The rest of them are shown in part only, being truncated by the circular margin of the map.

The meridians are composite curves converging on the pole in a way which is reminiscent of Bonne's. Of the meridians, only the central meridians are correct to scale. Other meridians are too long, increasing in length somewhat away from the central meridians. Moreover, the scale along each individual meridian increases slightly but progressively away from the pole.

Similarly, the length of the parallels is exaggerated away from the pole and the scale along each increases away from the central meridians. So, of all the meridians and parallels, the only ones which are correct to scale are the central meridians and they are the only ones to be equally divided. Only in this respect can the projection be termed 'equidistant'.

Because of the exaggeration of both meridian and parallel scales away from the pole and from the central meridians, the land-masses are progressively exaggerated in area and are, therefore, distorted in shape in those directions. The projection centred on London considerably exaggerates the area of Africa and that part of South America which can be shown. Both continents are elongated from east to west.

The great merits of the projection are that:

(*a*) the projection can be centred on any point on the earth's surface, and
(*b*) both bearing and distance from the centre of the projection are correct.

It is, therefore, valuable for use in any large international airport placed in the centre of the map because, from the centre, both bearing and distance to any part of the hemisphere shown are merely taken from the map. On pages 8 and 9 of Philips' *Modern School Atlas*, edited by Harold Fullard, six maps on this projection are shown centred on the following places: London, the antipodes of London, San Francisco, Cape Town, Cairo and Shanghai. They should be studied closely in conjunction with this description.

Graphical Construction of the Equatorial Zenithal Equidistant (*Fig.* 39)

(1) Draw a circle of radius $\pi r/2$.
(2) Draw in the equator as a diameter of the circle and, at right angles to it, the central meridian, also a diameter. Both are equal to πr, the correct length.
(3) Subdivide both into 12 equal parts, each $\pi r/12$, if meridians and parallels are to be shown at 15° intervals.
(4) Subdivide the bounding arc of each quadrant into 6 equal parts, by drawing angles of 15° at the centre starting from the equator in each case.
(5) The parallels are then inserted, as arcs of circles, through the corresponding subdivisions of the bounding arcs and the central meridian, i.e. through *L*, *A* and *T* (60°N. on Fig. 39).

The centre of such a circular parallel is a point, *X*, where the perpendicular bisector, *XY*, of *AT* meets the central meridian produced.

The other parallels are inserted in a similar way.

(6) The meridians (30°W. on Fig. 39) are similarly inserted by drawing an arc of a circle through *N.*, *E.* and *S.* The centre of this circular meridian is a point, *Q*, where the perpendicular bisector, *RQ*, of *ES* meets the equator produced; similarly for all other meridians.

15 Lambert's Polar Zenithal Equal-area (Fig. 40)

Equal-area in a polar zenithal projection is achieved by a suitable spacing of the parallels—in much the same way as the areas of the zones on Mollweide are made correct in area.

Because the spacing of the parallels is the sole distinguishing feature of this projection, it is desirable that readers should understand the simple mathematics involved in determining the correct spacing to attain the equal-area property. All parts of the projection must be made equal in area to their counterparts on the globe.

Step 1. Each zone on the projection must be made equal in area to the corresponding zone on the globe. The area of a zone on the globe $= 2\pi r.h$, that is:
The length of the equator × the perpendicular distance between the planes of the bounding parallels of the zone.
In Fig. 41 the area of the zone between the pole and latitude 60°N. $= 2\pi r \times NA$.
$NA = NO - AO$, i.e. $r - AO$.

In triangle TAO: $\dfrac{AO}{OT} = \sin \text{ angle } ATO$

$$AO = OT \times \sin \text{ angle } ATO$$
$$= r \sin 60°.$$

Therefore, $NA = r - r \sin \text{ latitude } 60°$.
Therefore, the area between parallel 60°N. and the pole $= 2\pi r\,(r - r \sin 60)$.

On the projection, this zone is to be represented as a circle. The problem now is to find the radius, x, of a circle which is equal in area to $2\pi r(r - r \sin 60)$:

$$\underset{\text{(Area of circle)}}{\pi x^2} = \underset{\text{(area of zone)}}{2\pi r\,(r - r \sin 60)}$$
$$x^2 = 2r\,(r - r \sin 60)$$
$$x = \sqrt{2r\,(r - r \sin 60)}$$

or, more simply $\sqrt{2r^2\,(1 - \sin 60)}$
Similarly for any other zone.

The limiting case on this projection is the zone between the equator and the pole, i.e. a hemisphere. The radius of the circle to represent the whole hemisphere is:
$\sqrt{2r^2(1 - \sin 0°)} = \sqrt{2r^2(1-0)} = \sqrt{2r^2} = 1{\cdot}414r$.

Step 2. On the globe, each zone is further divided into a number of equal parts by the equal spacing of the meridians. This is so on the projection because the meridians are spaced at equal and correct angular distances.

Description

(1) Parallels are concentric circles with the pole as centre and so spaced that the area between each of them is the same as that of the corresponding zone on the globe.
(2) Meridians are straight lines radiating from the pole at their true angular distances apart.
(3) The meridians and parallels intersect at right angles.
(4) The radial scale to ensure preservation of area is such that the parallels become progressively closer together equatorwards, i.e. the meridian scale diminishes progressively equatorwards but the diminution of scale is slight within 30° of the pole.

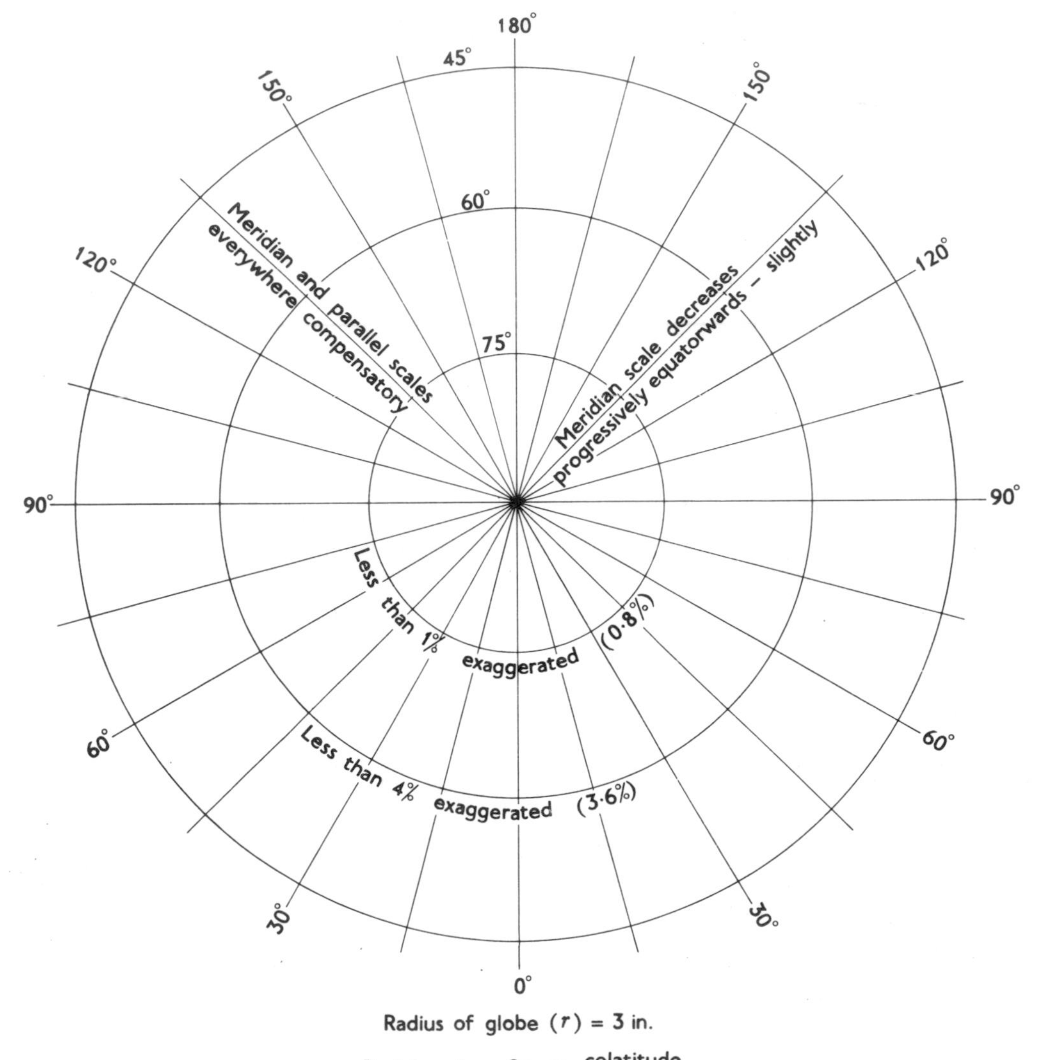

Radius of globe (r) = 3 in.

Radial scale = $2r \sin \frac{\text{colatitude}}{2}$

Fig. 40 Lambert's Polar Zenithal Equal-area.

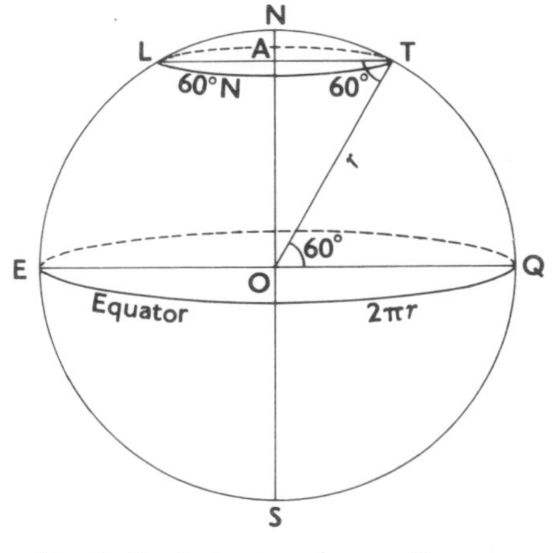

Fig. 41 To find area of zone between 60°N. and the North Pole.

(5) Because the meridians and parallels intersect at right angles and because the projection is equal-area, it follows that the meridian and parallel scales are compensatory—that is at any point the *diminution* of the meridian scale is balanced by an equivalent *exaggeration* of the parallel scale. At 75°N. the exaggeration of the parallel scale is 0·8%; at 60°N. it is 3·6%—still quite small. (Those interested in seeing why this is so should consult 'Calculation' at the end of the chapter.)

It follows that within 30° of the pole the shape of land-masses is quite well preserved, because there is only slight compression of them from north to south and only a slight corresponding stretching from east to west.

(6) Moreover, as on all zenithals, true direction from the centre of the projection is maintained.

Uses

The preservation of area as well as reasonable shape equatorwards to about 60° latitude makes the projection a good one for representing polar areas, but much depends on the purpose for which the map is required. For distributions it is clearly suitable; but a number of polar zenithals have much to recommend them for other purposes. The gnomonic, while having the disadvantage of greater exaggeration of both meridian and parallel scales is not equal-area, but any straight line drawn on it is a great circle. The equidistant preserves true distance as well as direction from the centre of the map.

Calculation

To calculate the exaggeration along the parallels

$$\text{Exaggeration} = \frac{\text{Length of parallel ON PROJECTION}}{\text{Length of parallel ON GLOBE}}$$

$$= \frac{2\pi\sqrt{2r^2(1-\sin \text{latitude})}}{2\pi r \cos \text{latitude}} = \frac{\sqrt{2r^2(1-\sin \text{latitude})}}{r \cos \text{latitude}}$$

Assume radius of globe to be 1 inch.

$$\text{Exaggeration at } 75°\text{N.} = \frac{\sqrt{2(1-0{\cdot}9659)}}{0{\cdot}2588} = \frac{\sqrt{2-(2\times 0{\cdot}9659)}}{0{\cdot}2588}$$

$$= \frac{\sqrt{0{\cdot}0682}}{0{\cdot}2588} = \frac{0{\cdot}2610}{0{\cdot}2588};$$

$$\text{Percentage exaggeration} = \frac{26{\cdot}10}{0{\cdot}2588} = 100{\cdot}8$$

i.e. 0·8%

Similarly for 60° and 45° latitudes.

16 Lambert's Equatorial Zenithal Equal-area

This projection bears a superficial similarity to the Equatorial Zenithal Equidistant described in Chapter 14; but there are significant differences between the two projections. In the equidistant projection, the meridians are equally spaced along the equator, and the parallels are equally spaced along the central meridian; in fact, all meridians are *equidistantly* spaced and so are all the parallels. However, in the equal-area projection, the meridians at the equator become progressively closer together towards the periphery, and the parallels along the central meridian become progressively closer together towards the poles. In this respect, compare the spacing of the *parallels* on the *Polar* Zenithal Equidistant with that on the *Polar* Zenithal Equal-area previously discussed.

As in all types of equatorial zenithal, the projection can cover only half the globe and the centre of the projection is any selected point *on the equator*. The complete projection is enclosed within a circle the area of which is equal to that of a hemisphere; the meridians and parallels are so arranged that every part of the projection is equal in area to its counterpart on the globe.

Construction (*see Fig.* 42)

Suppose (*a*) that the centre of the projection is the point where the Greenwich Meridian cuts the equator, and (*b*) that meridians and parallels are at 30° intervals.

(1) Select a suitable radius, r, of the globe to be represented.
(2) Draw a circle of radius $1{\cdot}414r$. This circle, which represents the bounding meridians 90°W. and 90°E., will enclose an area equal to a hemisphere.
(3) Draw in a diameter to represent the equator and, at right angles to it, another diameter to represent the central meridian (0°) so that they cut at G, the centre of the projection.
(4) At the centre, G, draw angles of 30° (see supposition (*b*) above) starting from the equator so that their arms cut the bounding meridians at A, B, C and D (and similarly for the southern hemisphere).
(5) (*a*) Mark off from G distances GW, GX, GY and GZ, all equal to $\sqrt{2r^2(1-\cos 60)}$ or $r\sqrt{2(1-\cos 60)}$. For explanation, see (2) under 'Calculation' at the end of this chapter. An arc of a circle drawn through B, X and C would then represent parallel 60°N.
(*b*) Draw, with centre G and radius $\sqrt{2r^2(1-\cos 30)}$, another circle which cuts the equator at H and J, and the central meridian at I and K. An arc of a circle drawn through points A, I and D would represent the parallel of latitude 30°N.
(6) Meridians 60°W. and 30°W. are inserted by drawing arcs of circles through N, W, S, and N, H, S respectively. Similarly, 60°E. and 30°E. are inserted by drawing arcs through N, Y, S and N, J, S respectively.

Description

(1) The projection can cover only half the globe.
(2) The two bounding meridians (180° apart) of the complete projection form a circle whose radius is exactly the same as that of the central circle on the conventional form of Mollweide—that is, $1{\cdot}414r$, where r is the radius of the

globe from which the projection is developed. It is also the same size as the equator in the polar case of the Zenithal Equal-area.

The central meridian is a straight line of length $1{\cdot}414r \times 2$. All other meridians are arcs of circles of progressively greater radius towards the centre; they all converge on the poles.

(3) Only half the equator is shown; it is a straight line at right angles to the central meridian and of the same length. The poles are represented as points. All other parallels are shown as arcs of circles of progressively smaller radii polewards. Clearly, they are not concentric.

(4) All parallels cut the central meridian at right angles and all meridians cut the equator at right angles. The intersections of all other meridians and parallels become progressively more acute away from the central meridian towards the periphery and more so in higher latitudes than in lower.

(5) The distance from the centre to the point of intersection of any meridian with the equator is $\sqrt{2r^2(1-\cos x^\circ)}$, where x° is the longitudinal distance of the meridian from the centre, that is:

$\sqrt{2r^2(1-\cos 30)}$, $\sqrt{2r^2(1-\cos 60)}$, etc.

The points of intersection of each successive meridian with the equator become progressively closer together towards the periphery; in other words, the equatorial scale decreases progressively away from the centre of the projection. In fact, relative to the length of the equator on the globe, there is a progressive diminution of the scale along it away from the centre. The following table shows the extent of the shortening of the equator at various points along it:

Distance in degrees from the centre of the projection	*Percentage diminution of the equatorial scale*
10°	0·32
20°	0·83
30°	1·07
40°	2·05
60°	4·52
90°	10·01

Those interested should consult 4(*a*) to (*c*) under 'Calculation' below.

The other parallels, although in total length shorter than they should be, are not greatly diminished. Within 40° of the centre, the diminution of scale along them is very small.

(6) The scale characteristics of the central meridian are the same as those along the equator; the parallels cut it at progressively shorter intervals towards the poles. Each successive meridian increases in length towards the margins of

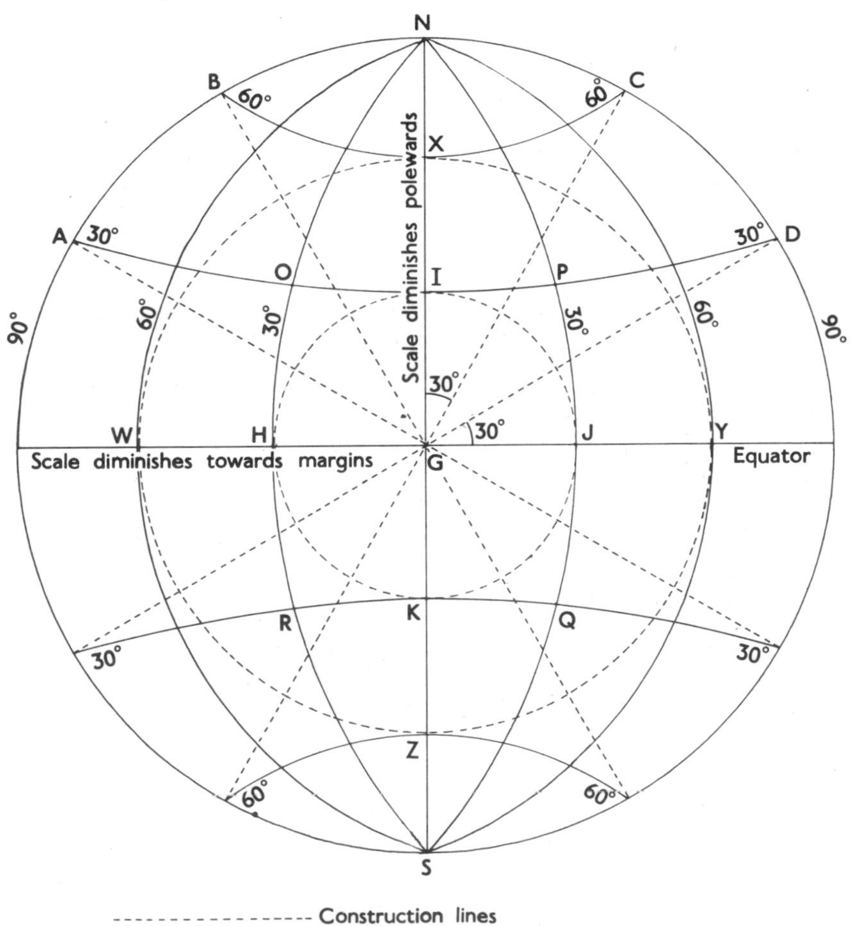

Fig. 42 Lambert's Equatorial Zenithal Equal-area.

the projection. The central meridian is 10% too short; one meridian on either side of the central meridian is the correct length, but thereafter they become progressively too long until the bounding meridians are 41% exaggerated. (See (5) under 'Calculation' below.)

(7) Towards the margins of the projection, the shape of land-masses is badly distorted:

(*a*) They are 'pulled out of upright' by the obliquity of the meridians to the parallels in middle and high latitudes especially towards the margins of the map.
(*b*) They become increasingly compressed longitudinally because of the progressive diminution of scale along each parallel away from the central meridian.
(*c*) Towards the margins, they become correspondingly elongated to compensate for this increasing compression.

However, within 40° of the centre of the projection along both the central meridian and the equator, the amount of compression is so small that, for most practical purposes, it can be ignored.

(8) The projection is equal-area because the scales along both meridians and parallels are diminished away from the centre to such an extent as to eliminate the effect of the earth's curvature.

(9) Direction from the centre of the projection is correct; any straight line drawn from the centre of the projection is a line of true direction or bearing.

Uses

As a projection for a whole hemisphere, the Equatorial Zenithal Equal-area is of very little value because the longitudinal compression and latitudinal elongation of land-masses virtually destroys their shape; but, within 40° of the centre, the error of scale along both meridians and parallels *and* the obliquity of their intersection are so small that it gives rise to a very good map of any area which is more or less equally divided by the equator and lies within this limit.

Africa, which extends between 18°W. and 50°E. and between 35°S. and 37°N., is very well represented on it.

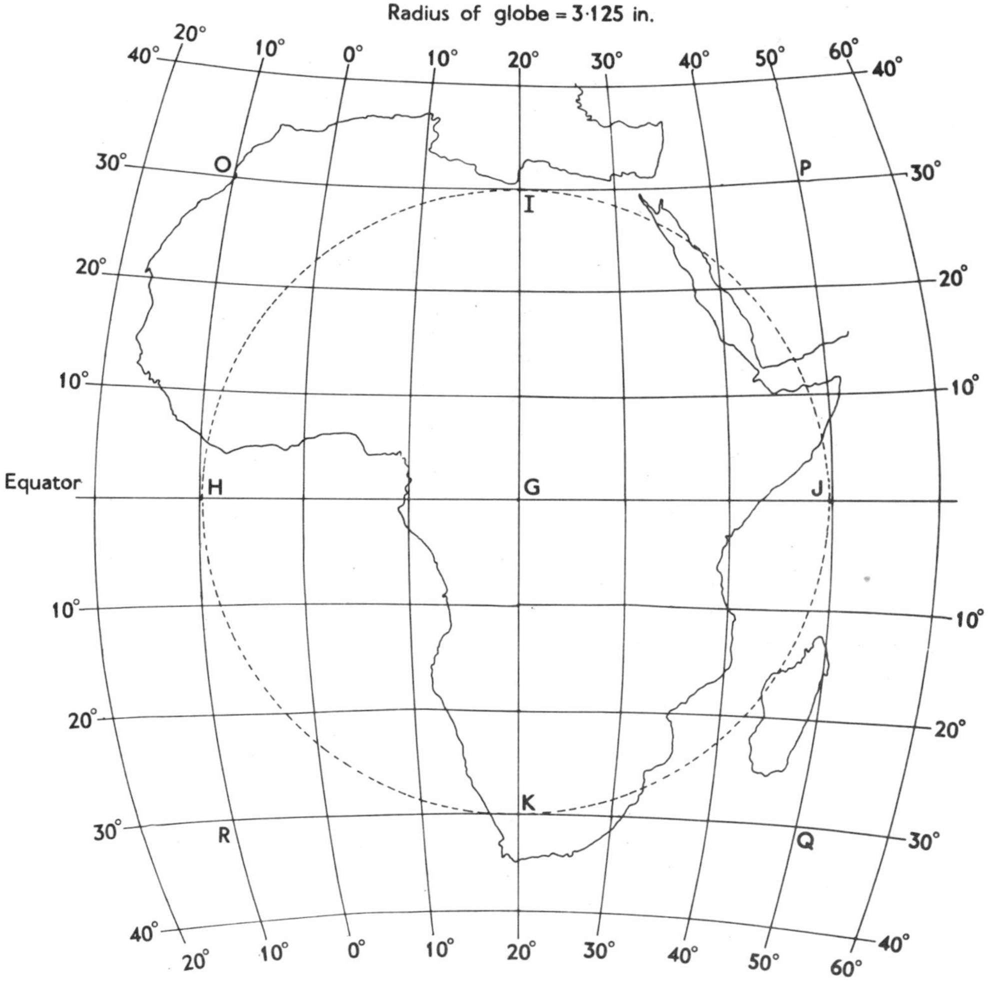

Fig. 43 Africa on the Equatorial Zenithal Equal-area.
(1) Area is preserved.
(2) Within 40° of the centre of the map, the error of both parallel and meridian scales is small; hence shape is well represented.
(3) Any straight line drawn from the centre of the map is a line of true direction or bearing.
OPQR covers the same region as *OPQR* in Fig. 42.

(*a*) Its area is preserved. (*b*) Shape is very well represented. (*c*) Direction from the centre of the map is correct. The only disadvantage is that, unlike Sanson–Flamsteed, the projection does not have a uniformity of scale along either meridian or parallel, but even so the error along both is minimal.

It could similarly be used with advantage for other areas virtually bisected by the equator and within 40° of it in any direction, for example, South-east Asia with Australia, *or* Central America, the Caribbean and South America (to include the whole of Brazil). It is used in Philips' *Modern School Atlas* and the *University Atlas* for maps of the whole of Africa.

Calculation

If the radius of the globe is 3 inches, and the radius of the bounding meridians (90°W. and 90°E.) is x, then:

(1) area enclosed by them (πx^2) must be equal to a hemisphere ($2\pi r^2$).

$\pi x^2 = 2\pi r^2$. Therefore, $x^2 = 2r^2$.

$x = \sqrt{2}r = 1{\cdot}414r$

$= 1{\cdot}414 \times 3 \quad = 4{\cdot}242$ inches.

(2) *To show that the spacing of the meridians along the equator and of the parallels along the central meridian is:*

$$\sqrt{2r^2(1-\cos \text{latitude/longitude})}$$

Suppose the centre of the proposed projection is the point of intersection of meridian 0° with the equator (*G* on Fig. 42). Using a compass, with this point as centre, *draw on the globe* a circle to pass through the point where meridian 60°W. cuts the equator (a point represented by *W* on Fig. 42). Label this circle, *Circle E*. It would also pass through the intersections of 60°N. and 0° longitude, 60°W. and the equator, and 60°S. and 0° longitude. The area of the surface of the globe enclosed by Circle E would be the same as *the surface area, on the globe, of the zone between* 30°*N. and the pole.* (Examine the globe to establish this for yourself.)

In the *polar* case of the Zenithal Equal-area, the problem was to discover the radius of a circle (to represent, for example, 30°N.) which would equal in area the zone between 30°N. and the pole. In Chapter 15, Step 1, this was shown to be $\sqrt{2r^2(1-\sin \text{latitude})}$, in this case, $\sqrt{2r^2(1-\sin 30)}$. But sin 30 is equal to cos 60. Hence the radius of this circle could quite readily be expressed as $\sqrt{2r^2(1-\cos 60)}$. This would also be the radius of a circle, *WXYZ* on Fig. 42, which could be drawn on the *equatorial* case of the projection to equal in area the portion of the surface of the globe enclosed by Circle E above. It should be clear, therefore, that the distance from the centre of the projection to the point where any meridian cuts the equator is $\sqrt{2r^2(1-\cos \theta^\circ)}$, where θ° is the difference in longitude between the centre of the projection and the meridian concerned. (Similarly, the distance from the centre of the projection to the point where any selected parallel cuts the central meridian is $\sqrt{2r^2(1-\cos \lambda^\circ)}$, where λ° is the latitude of the parallel concerned.)

It is seen, therefore, that it is mathematically fairly simple to fix the positions of points *H*, *I*, *J* and *K* in relation to *G*. It is more difficult to fix mathematically the points *O* (30°N., 30°W.), *P* (30°N., 30°E.), *Q* (30°S., 30°E.) and *R* (30°W., 30°S.). It is also, for most purposes, unnecessary because the meridians and parallels which intersect at these points can be drawn as circles quite easily by graphical methods used in the above construction, paras. (5) and (6).

(3) Distance of meridians along the equator from the centre of the projection = Distance of parallels along the central meridian from the centre of the projection = $r\sqrt{2(1-\cos \text{latitude or longitude.})}$

(*a*) 30°W. $r\sqrt{2(1-\cos 30)} = 3\sqrt{2(1-0{\cdot}866)}$
$= 3\sqrt{2\times 0{\cdot}134} \quad = 3\sqrt{0{\cdot}268}$
$= 3\times 0{\cdot}518 \quad = 1{\cdot}554$ inches.

(*b*) 60°W. $r\sqrt{2(1-\cos 60)} = 3\sqrt{2(1-0{\cdot}5)}$
$= 3\sqrt{(2\times 0{\cdot}5)} \quad = 3\sqrt{1} \quad = 3$ inches.

(4) Diminution of scale along the central meridian and equator:

(*a*) 30° *from the centre of the projection*

$$\frac{\text{Distance ON PROJECTION}}{\text{Distance ON GLOBE}} = \frac{\sqrt{2r^2(1-\cos 30)}}{\pi r/6} =$$

$$\frac{6\sqrt{2r^2(1-\cos 30)}}{\pi r} = \frac{6r\sqrt{2(1-\cos 30)}}{\pi r} =$$

$$\frac{6\sqrt{2(1-\cos 30)}}{\pi} = \frac{6\sqrt{2(1-0{\cdot}866)}}{3{\cdot}142} =$$

$$\frac{6\sqrt{0{\cdot}268}}{3{\cdot}142} = \frac{6\times 0{\cdot}518}{3{\cdot}142} = \frac{3{\cdot}108}{3{\cdot}142} = 0{\cdot}9893$$

Percentage diminution $= 100 - 98{\cdot}93 = 1{\cdot}07\%$

(*b*) 60° *from the centre of the projection*

$$\frac{\text{Distance ON PROJECTION}}{\text{Distance ON GLOBE}} = \frac{\sqrt{2r^2(1-\cos 60)}}{\pi r/3} =$$

$$\frac{3\sqrt{2r^2(1-\cos 60)}}{\pi r} = \frac{3r\sqrt{2(1-\cos 60)}}{\pi r} =$$

$$\frac{3\sqrt{2(1-\cos 60)}}{\pi} = \frac{3\sqrt{2(1-0{\cdot}5)}}{3{\cdot}142} =$$

$$\frac{3\sqrt{1}}{3{\cdot}142} = \frac{3{\cdot}0}{3{\cdot}142} = 0{\cdot}9548$$

Percentage diminution $= 100 - 95{\cdot}48 = 4{\cdot}52\%$

(*c*) 90° *from the centre of the projection*

$$\frac{\text{Distance ON PROJECTION}}{\text{Distance ON GLOBE}} = \frac{\sqrt{2r^2(1-\cos 90)}}{\pi r/2} =$$

$$\frac{2\sqrt{2r^2(1-\cos 90)}}{\pi r} = \frac{2r\sqrt{2(1-\cos 90)}}{\pi r} =$$

$$\frac{2\sqrt{2(1-\cos 90)}}{\pi} = \frac{2\sqrt{2\times 1}}{3{\cdot}142} =$$

$$\frac{2\times 1{\cdot}414}{3{\cdot}142} = \frac{2{\cdot}828}{3{\cdot}142} = 0{\cdot}8999$$

Percentage diminution $= 100 - 89{\cdot}99 = 10{\cdot}01\%$

(5) *Exaggeration of the length of the bounding meridians* (90°W. and 90°E.)

$$\frac{\text{Length ON PROJECTION}}{\text{Length ON GLOBE}} = \frac{\pi x}{\pi r}\times 100, \text{ where } x = 1{\cdot}414r$$

$$\frac{\pi\times 1{\cdot}414r}{\pi r}\times 100 = 1{\cdot}414\times 100 = 141{\cdot}4\%$$

i.e. 41·4% exaggerated.

17 Lambert's Oblique Zenithal Equal-area

As in the oblique case of the Zenithal Equidistant (Chapter 14), the Oblique Zenithal Equal-area can represent only half the globe. In its complete form (Fig. 44), the margins are a complete circle whose radius is $1{\cdot}414r$, where r is the radius of the globe. This is the same as the radius of the circle forming the pair of marginal meridians in the *equatorial* case of the Zenithal Equal-area considered in the previous chapter. The centre of the projection can be any point between the equator and either pole.

The vertical diameter is again composed of portions of two meridians 180° apart. If the projection is centred on a point at longitude 0°, latitude 45°N., the central meridian will consist mainly of the Greenwich Meridian from the pole southwards to 45°S. and the rest of it will represent meridian 180° from the pole to 45°N. It will be remembered that in the *Polar* Zenithal Equal-area (Chapter 15), the distance of any parallel (along any meridian) from the pole was shown to be $\sqrt{2r^2(1-\sin \text{latitude})}$ and that, because of this, the parallels became closer together with increasing distance from the pole. The same formula is used for sub-dividing the central meridian in the *oblique* case as a first step in making the projection equal-area.

Some of the parallels, for example, 60°N. (on Fig. 44) are shown completely but they are composite curves rather than arcs of circles; the remainder are shown in part only. The meridians, too, are composite curves which converge on the pole.

The scale along the central meridian is diminished, away from the centre of the projection, to the same extent as along the equator *and* the central meridian on the equatorial case (see table on page 55). The percentage diminution varies between 0·32, 2·0 and 10·0 at points along it which are respectively 10°, 40° and 90° from the centre. The other meridians are longer than the central meridian and consequently the amount by which they are diminished is smaller.

The length of the parallels is exaggerated away from the

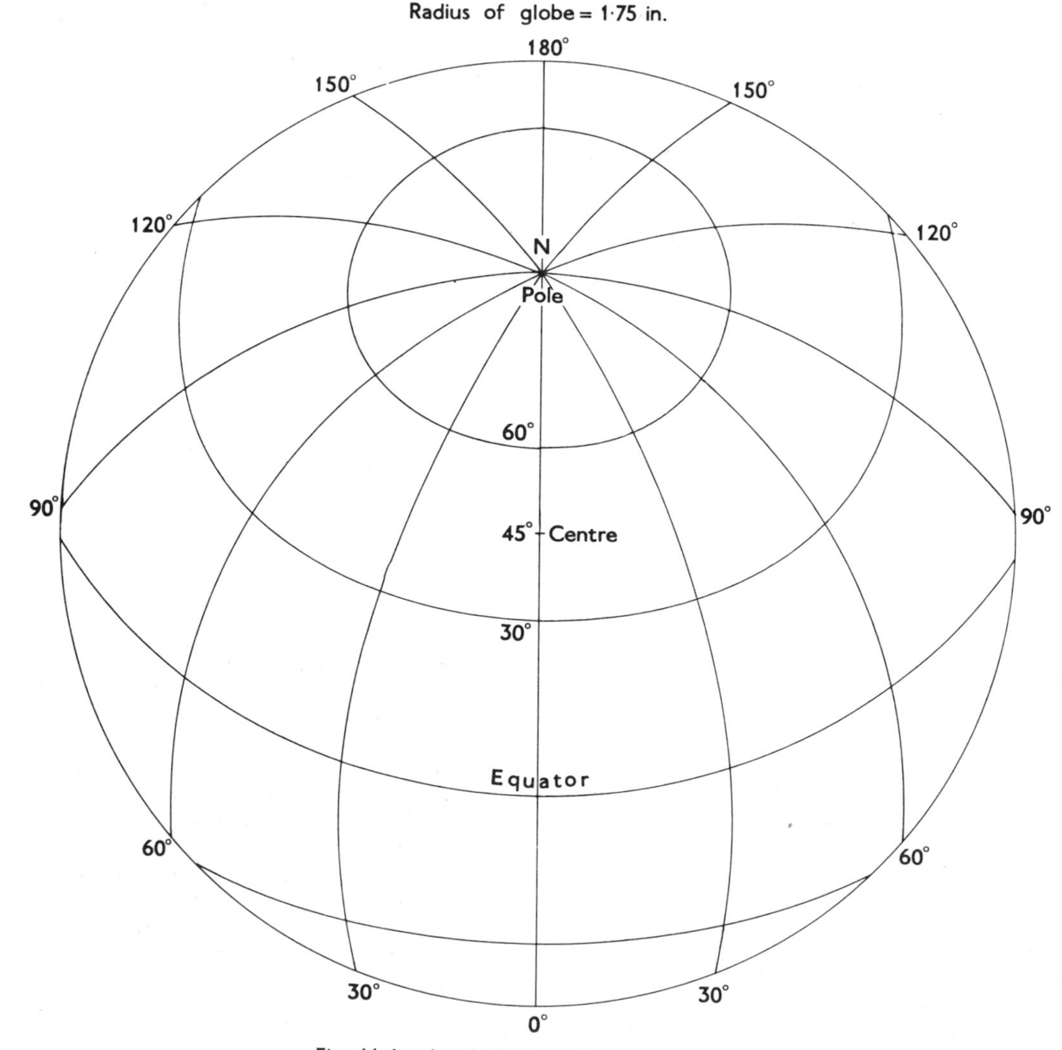

Fig. 44 Lambert's Oblique Zenithal Equal-area.

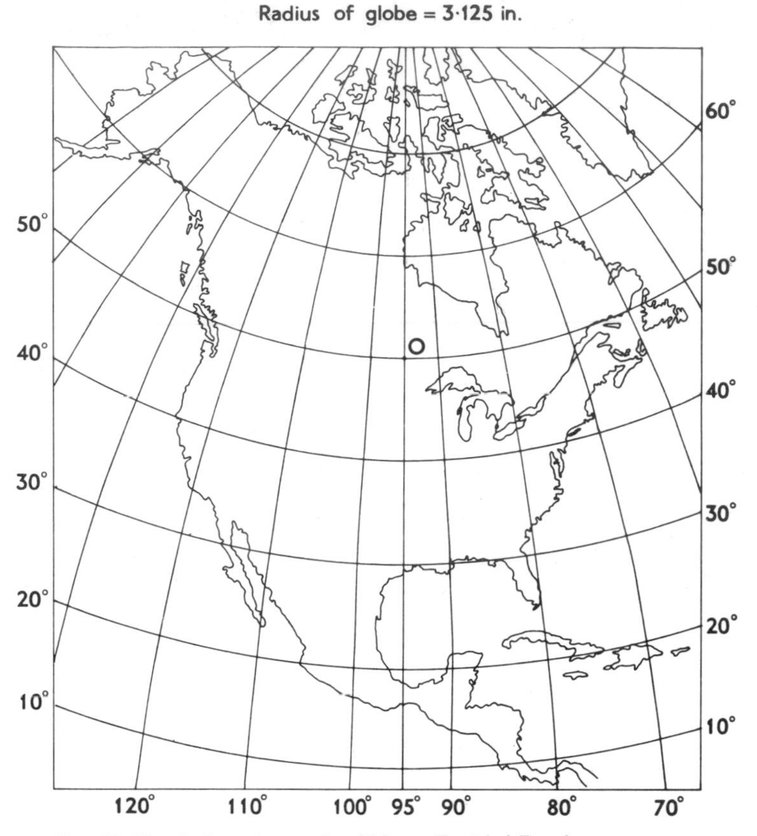

Fig. 45 North America on the Oblique Zenithal Equal-area.
Point *O* is the centre of the map, i.e. point of contact of plane with the globe.
(1) Area is preserved.
(2) Within 40° of the centre, the error of parallel and meridian scales is small, and the intersection of parallel with meridian is near right-angled. Hence shape is well represented.
(3) Any straight line drawn from the centre of the map is a line of true direction or bearing.

centre of the projection to compensate for the diminution of the meridian scale. Away from the centre of the map, land-masses are increasingly compressed latitudinally and correspondingly elongated longitudinally. However, as in the equatorial case, the amount of compression—about 2% —within 40° of the centre of the map is so small that in practice it can virtually be ignored.

In common with all zenithal projections, it has the valuable property that any straight line drawn from the centre of the map is a line of true bearing.

Uses

For representing a whole hemisphere, the Oblique Zenithal Equal-area is of little value because latitudinal compression and longitudinal elongation of land-masses results in poor representation of shape. However, within 40°, or even more, from the centre, the error of scale along meridians and parallels and the obliquity of their intersection are so small that distortion of shape is minimal. It is very suitable for any continent—wholly in one hemisphere—which is limited to about 40° from the centre of the map in any direction. North America (see Fig. 45) which, including the Canadian archipelago, extends from about 80°N. to 27°N. and from 160°W. to 60°W., is well represented on it if the centre of the projection is at 50°N. and 95°W. or 100°W. It has the following advantages: (*a*) its area is preserved; (*b*) its shape is very well represented; (*c*) from the centre of the map direction is correctly represented by a straight line.

Compared with Bonne, however, it has the disadvantage of not possessing accuracy of scale along either meridian or parallel. Bonne represents the shape of North America equally well, it preserves area and maintains accuracy of scale along the central meridian and all parallels, but true direction cannot be represented on it by a straight line from the centre of the map. The Oblique Zenithal Equal-area gives an equally good result for Europe, South America and, to a lesser extent, for Asia.

PART VI

18 Choosing a Map Projection

The choice of a map projection depends entirely on the purpose for which the map is required.

Distributions

Maps to show distributions demand an equal-area projection to make regional comparisons of area possible. For distributions of various kinds within the Tropics, Lambert's Cylindrical Equal-area could be used as the exaggeration of the parallel scale at the Tropics and the corresponding diminution of the meridian scale is only about 9%. The shape of land-masses is not seriously affected by the very slight north–south compression and east–west elongation which the projection imposes on them. Sanson–Flamsteed, another equal-area projection, could be used equally well for the same purpose. Again, the distortion of shape is slight within the Tropics, but it is of a different kind; in this case, it stems from the increasing obliquity of meridians to the parallels and the consequent slight but progressive elongation of the meridians towards the margins. Mollweide has a similar effect on the shape of land-masses, so there is very little to choose between any of these three for equal-area maps of tropical areas.

However, for distribution maps showing the world on one sheet, the Cylindrical Equal-area is quite unsatisfactory because the north–south compression and east–west elongation of land-masses makes them virtually unrecognisable in high latitudes. The choice then lies between Sanson–Flamsteed and Mollweide which both distort shape in somewhat the same way in middle and high latitudes towards the margins of the map; but the shape of land-masses is rather better preserved on Mollweide because of the shortening of the length of the equator and the varying scale along the parallels; together, these characteristics contribute towards a better shape by reducing the acuteness of the angle of intersection of meridian and parallel. If the purpose of the map permits, both projections are considerably improved by interruption over the oceans. (See Chapter 11.)

General-purpose Maps of the World

Mollweide and Sanson–Flamsteed are suitable, especially if interrupted, for the reasons already given. Mercator, however, is quite unsatisfactory because of the gross exaggeration of area and corresponding distortion of shape of large continental areas which have a great extent in latitude. It can be used only for special purposes, especially those which require the representation of true direction, such as winds and ocean currents. It is, of course, invaluable for navigation because on it any straight line is a line of constant bearing. Any of the Zenithal Gnomonics, the polar, the oblique or equatorial cases, should be used in conjunction with it because, the use of tables excepted, they provide the simplest means of plotting great circle courses on Mercator; on all Zenithal Gnomonics, any straight line is a great circle.

Maps of Individual Continents

If preservation of area is required for a continent wholly in one hemisphere but with a relatively small extent in longitude, for example, for North America, Australia, Europe or—marginally—Asia, Bonne's is suitable; but the great longitudinal extent of Eurasia is far too large for Bonne. Lambert's Zenithal Equal-area (the oblique case) is equally suitable and has the added advantage that, in common with all other zenithal projections, bearings from the centre of the map are true.

Representation of Africa

Africa, and to some extent South America, is in a special category in so far as it lies astride the equator and its latitudinal extent is almost equally divided by it; at the same time, its longitudinal extent (20°W. to 50°E.) is not so great as to suffer any great marginal distortion of shape if it is superimposed on a Sanson–Flamsteed (equal-area) projection. The acuteness of the intersection of the meridians with the parallels is not very great at the margins.

Mollweide produces a map of Africa which is equally good in preserving area with minimal distortion of shape, but it has the disadvantage, not suffered by Sanson, of lack of uniformity of scale along the parallels and along the central meridian so that distances cannot be taken directly from the map along these lines. Mollweide is also more difficult to construct and possesses no overriding advantage in some other respect. This cannot be said of Lambert's Zenithal Equal-area (the equatorial case); like Mollweide, it is more difficult to construct than Sanson and, like it, does not possess a degree of scale uniformity, but it *does* have property which could be an overriding advantage and that is that direction from the centre of the map is correct. South America is similar to Africa. South America is similar to Africa except in two significant ways:

(*a*) it is not evenly divided by the equator, and
(*b*) its longitudinal extent is smaller—about 45° at its broadest compared with 70° in the case of Africa.

Sanson–Flamsteed, besides preserving accuracy of area, presents its shape reasonably well when 60°W. is taken as its central meridian. So does Lambert's Zenithal Equal-area which is widely used for it.

Maps of Small Countries

Maps of small countries, especially those with a small extent in latitude rather than in longitude are very well represented on the Conical Projection with Two Standard Parallels. It preserves neither shape nor area accurately but does both with minimal error provided the latitudinal extent is no more than about 15°. Any individual country of Europe, even Scandinavia, can be reasonably well represented. Bonne's, of course, preserves shape equally well if the longitudinal extent is restricted, and it has the added advantage of being equal-area.

The Relative Value of Polar Zenithals

For representing polar areas, there is not a great deal of difference between a number of polar zenithals—the Equidistant, the Equal-area and the Stereographic, provided they do not extend beyond 60° latitude, i.e. 30° from the pole. The differences in radial scale are quite small and so, therefore, are the differences in scale along the parallels. Moreover, each has a valuable property—equality of area or preservation of scale from the centre of the map, or orthomorphism. On the Polar Zenithal Gnomonic, however, the radial scale increases more rapidly up to 60° latitude and even more rapidly equatorwards of this so that shape is not so well preserved nor is area; but, if its particular property, that of representing a great circle by a straight line is required above all else, then associated disadvantages such as poorer representation of shape and exaggeration of area equatorwards must be accepted.

The choice of map projection is clearly then a matter of balancing advantage against disadvantage for any given purpose; a perfect solution can never be found for this is the inherent problem, but some solutions are better than others. The understanding of the principles of map projection can help one to make a logical choice.

Index

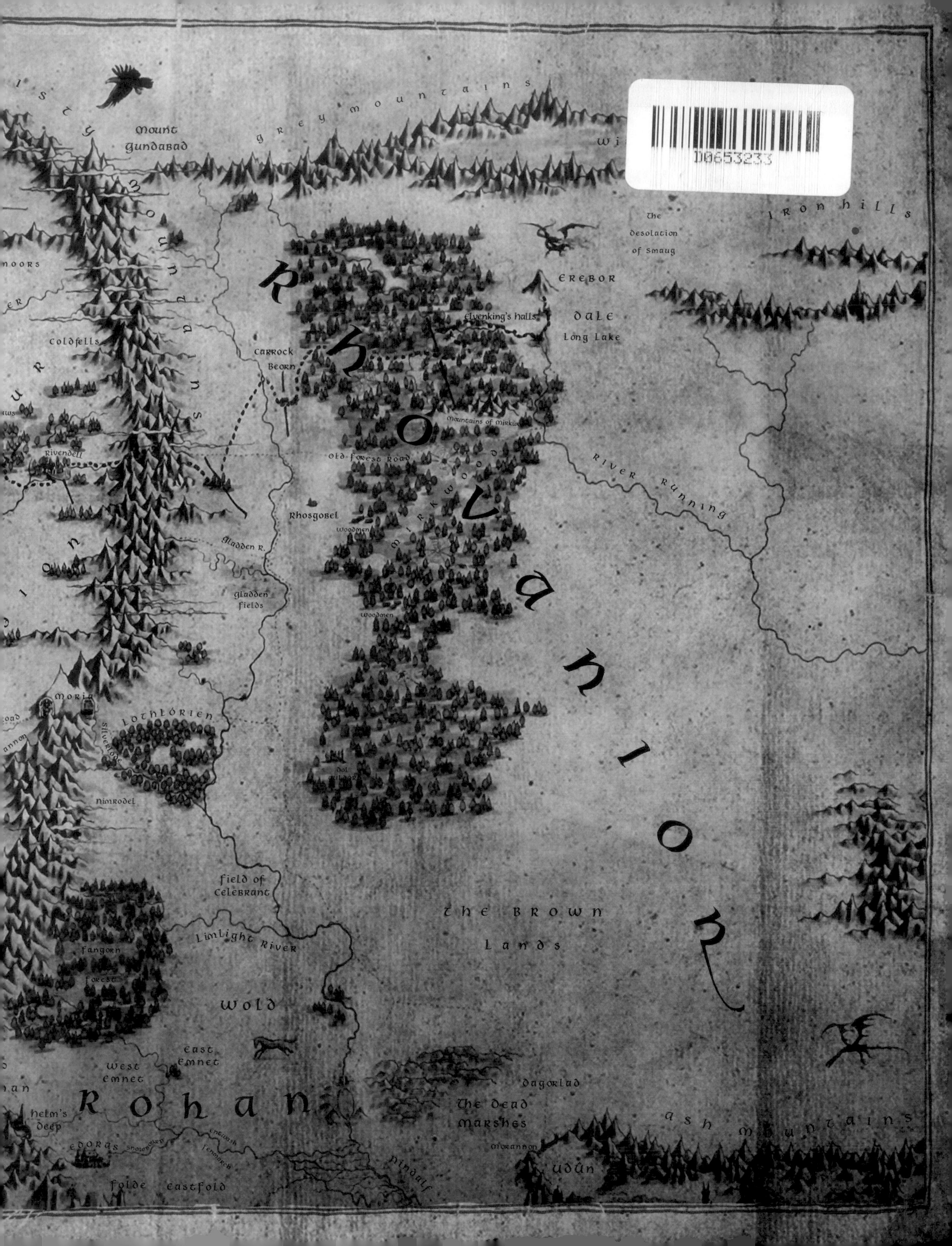
Grey Mountains
Mount Gundabad
Misty Mountains
Iron Hills
The desolation of Smaug
Erebor
Elvenking's halls
Dale
Long Lake
Carrock
Beorn
Coldfells
Rhovanion
Mountains of Mirkwood
Old Forest Road
Rivendell
River Running
Rhosgobel
Woodmen
Mirkwood
Gladden R.
Gladden Fields
Moria
Lothlórien
Silverlode
Nimrodel
Field of Celebrant
The Brown Lands
Limlight River
Fangorn Forest
Wold
East Emnet
West Emnet
Rohan
Dagorlad
The Dead Marshes
Ash Mountains
Helm's Deep
Edoras
Udûn
Nindalf
Eastfold
D0653233

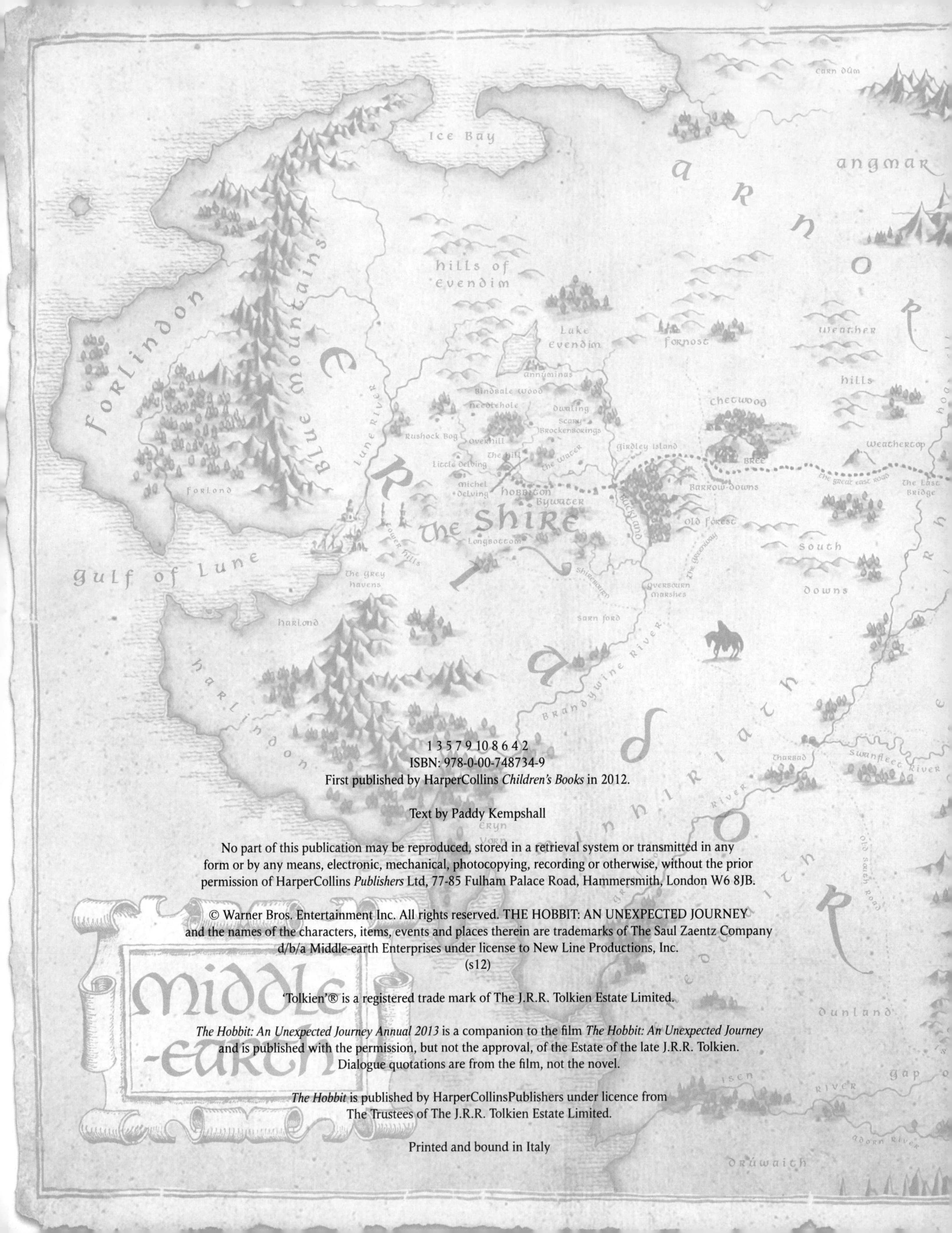

1 3 5 7 9 10 8 6 4 2
ISBN: 978-0-00-748734-9
First published by HarperCollins *Children's Books* in 2012.

Text by Paddy Kempshall

The Hobbit: An Unexpected Journey Annual 2013 is a companion to the film *The Hobbit: An Unexpected Journey* and is published with the permission, but not the approval, of the Estate of the late J.R.R. Tolkien. Dialogue quotations are from the film, not the novel.

Printed and bound in Italy

THE HOBBIT™
AN UNEXPECTED JOURNEY
Annual 2013
HarperCollins Children's Books

CONTENTS

Bilbo Baggins

Like all hobbits, Bilbo is fond of his comfortable life in the Shire. However, when the Wizard, Gandalf, and 13 Dwarves unexpectedly appear on his doorstep, his Tookish spirit of adventure gets the better of his Baggins good sense and changes his life forever.

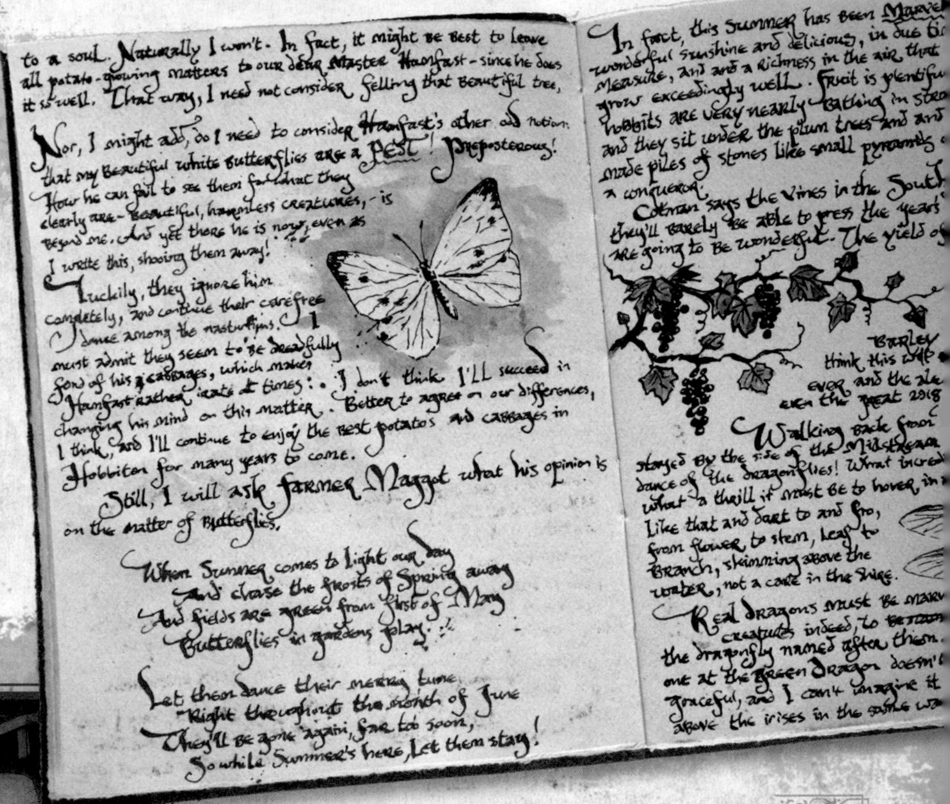

Bilbo likes to keep a journal, and his adventures with the Company of Dwarves mean he certainly has a lot to write about!

Bilbo's home is a very fine hobbit hole called Bag End. With a full larder and a cheery fire, it's as warm and cosy a place as you could imagine.

Good hobbits always have something tasty cooking in the oven. Bilbo was certainly glad he had lots of food in the house when Thorin and his 12 hungry companions appeared one afternoon and decided to stay for dinner!

When Bilbo agrees to join Thorin Oakenshield, little does he know that a plain gold ring he finds along the way will be more precious than all of the treasure in the whole of Middle-earth.

On his adventures, Bilbo finds Sting – an Elven weapon made in the ancient city of Gondolin. It was used in the great Goblin wars and glows blue whenever Orcs and Goblins are near!

Bilbo is perhaps the most unexpected choice to join a band of heroes on a quest to steal a dragon's treasure, but along the way it is his bravery and skill that save everyone from disaster time and time again.

DID YOU KNOW?

Hobbits' large, hairy feet are so tough they can walk anywhere without wearing shoes or socks!

PANTRY PUZZLE

Look closely and see if you can spot the 8 differences between these two pictures of Bilbo and his visitors.

A.

B.

COMPANY CONTRACT

Thorin has heard of your deeds and wants you to join the Quest for the Lonely Mountain, just like Bilbo! If you think you have the skills, sign the contract as Burglar and begin your adventure.

In role as

Burglar

for Thorin and Company,

or in any other role they see fit,

at their sole discretion

from time to time.

Signed: Thorin son of Thrain

Witnessed: Balin son of Fundin

Burglar:

THE UNEXPECTED JOURNEY

When Bilbo agreed to join Thorin Oakenshield on his Quest to the Lonely Mountain, he had no idea of the adventures ahead of him. Follow the Company's path on this map and see where Bilbo's travels will take him.

RIVENDELL

The Elf-Haven of Imladris. A magical valley which is the home of Elrond and the Last Homely House East of the Sea. Hidden amongst the shadows of the Misty Mountains, Rivendell is a peaceful place where Bilbo and his friends can rest before continuing their quest.

THE SHIRE

The home of hobbits, the Shire is a simple land of rolling fields, good food and friendly folk. Bilbo lives here in a hobbit-hole just north of the large city of Hobbiton. It is here that his adventure with Thorin and his companions begins.

EREBOR
The Lonely Mountain – once it was the ancient palace of the Dwarven Kings under the Mountain, but now it is the home of the dragon, Smaug the Terrible. The final destination for Thorin and his companions, the Lonely Mountain's depths contain riches untold and dangers beyond imagination.
HALL OF THE ELVEN KING
Home of Thranduil, the Elven King. This home of the Woodland Elves could well be the end of the line for Bilbo and his companions. Luckily for the Company of Dwarves, Bilbo has a cunning plan and a magic ring that might just save the day...
MIRKWOOD
Once a glorious forest, full of life, Mirkwood is now a place of gloom, decay and evil. With deadly river waters that can send you into a deep sleep and creepy monsters that lurk in the shadows, Mirkwood is not a place for the faint of heart.
THE MISTY MOUNTAINS
A vast mountain range full of danger. Below them are huge caverns filled with Goblins and the terrible abode of the Goblin King – Goblin Town. It is no less dangerous above the ground in the Misty Mountains either. If the raging winds and storms don't sweep you from the path, then a boulder thrown by a Stone Giant just might!
Grey Mountains
Mount Gundabad
Desolation of Smaug
Erebor
Elvenking's halls
Dale
Long Lake
Coldfells
Carrock
Beorn
Rivendell
Old Forest Road
Rhosgobel
Woodmen
Gladden R.
Gladden Fields
Moria
Lothlórien
Silverlode
Nimrodel
Dol Guldur
Field of Celebrant
Limlight River
Fangorn
The Brown Lands
Ash Mountains
Helm's Deep
Edoras
Snowbourn
Folde
Eastfold
Nindalf
Morannon
Udûn

Gandalf

One of the most powerful Wizards in all of Middle-earth, Gandalf is wise and cunning. He has a soft spot for hobbits and sees that Bilbo's many hidden talents and strengths make him the perfect choice to help Thorin Oakenshield on his quest.

There is more to Gandalf than meets the eye. One of the ancient defenders of Middle-earth, it is his purpose to defend the world from evil.

On the way to the Lonely Mountain, Gandalf hears news that an ancient evil is returning to Middle-earth – a terrible enemy that he must try and stop.

The Elves are old friends of Gandalf and even have another name for him: Mithrandir.

When Elrond reads the runes on Gandalf's sword, Glamdring, he tells him that it was once the sword of an ancient and powerful King.

Gandalf is a member of the White Council. Made up of great Wizards and other magical beings, it is an ancient group that forever works to keep evil in check and Middle-earth safe.

A master Wizard, Gandalf is particularly skilled in conjuring and controlling fire in many different forms. From blinding flashes, to flaming explosions, Gandalf proves time and again that his magical powers are a match for any enemy.

Gandalf has visited the Shire many times over the years. He met Bilbo and his mother when Bilbo was a young boy – perhaps even then Gandalf saw hidden strengths in this tiny hobbit?

DID YOU KNOW?

Hobbits know Gandalf as someone who makes particularly amazing fireworks or 'whizz-poppers' for their parties!

BILBO'S BOTTLES

The Dwarves are thirsty! All of these bottles look the same, but only one is a bottle of Bilbo's finest nettle wine. Can you spot the bottle which is slightly different? That's the one that contains the wine!

Bilbo Baggins

Elrond

Elrond is one of the oldest and wisest Elves who still remain in Middle-earth, with a great ability to read many kinds of ancient runes. He is also a great fighter and played a key part in the final battle against the Dark Lord, Sauron, many years in the past.

As Master of the Last Homely House in Rivendell, Elrond welcomes Bilbo and the Company of Dwarves, allowing them to rest before continuing their quest into the Misty Mountains.

As part of the White Council alongside Gandalf, Elrond knows that the success of Thorin's mission is more important than many people suspect. In fact it is a crucial part of another plan to stop a far greater evil than Smaug from threatening Middle-earth.

It was Elrond who told Gandalf and Thorin the truth about the magic swords they found. Elrond also discovered the hidden moon runes on Thorin's map which revealed a secret entrance into the Lonely Mountain!

MAGIC SQUARES

Can you help Gandalf solve Elrond's tricky number puzzle? Fill in all the empty squares so that every row, column and small 3x3 square contains all the numbers from 1 to 9.

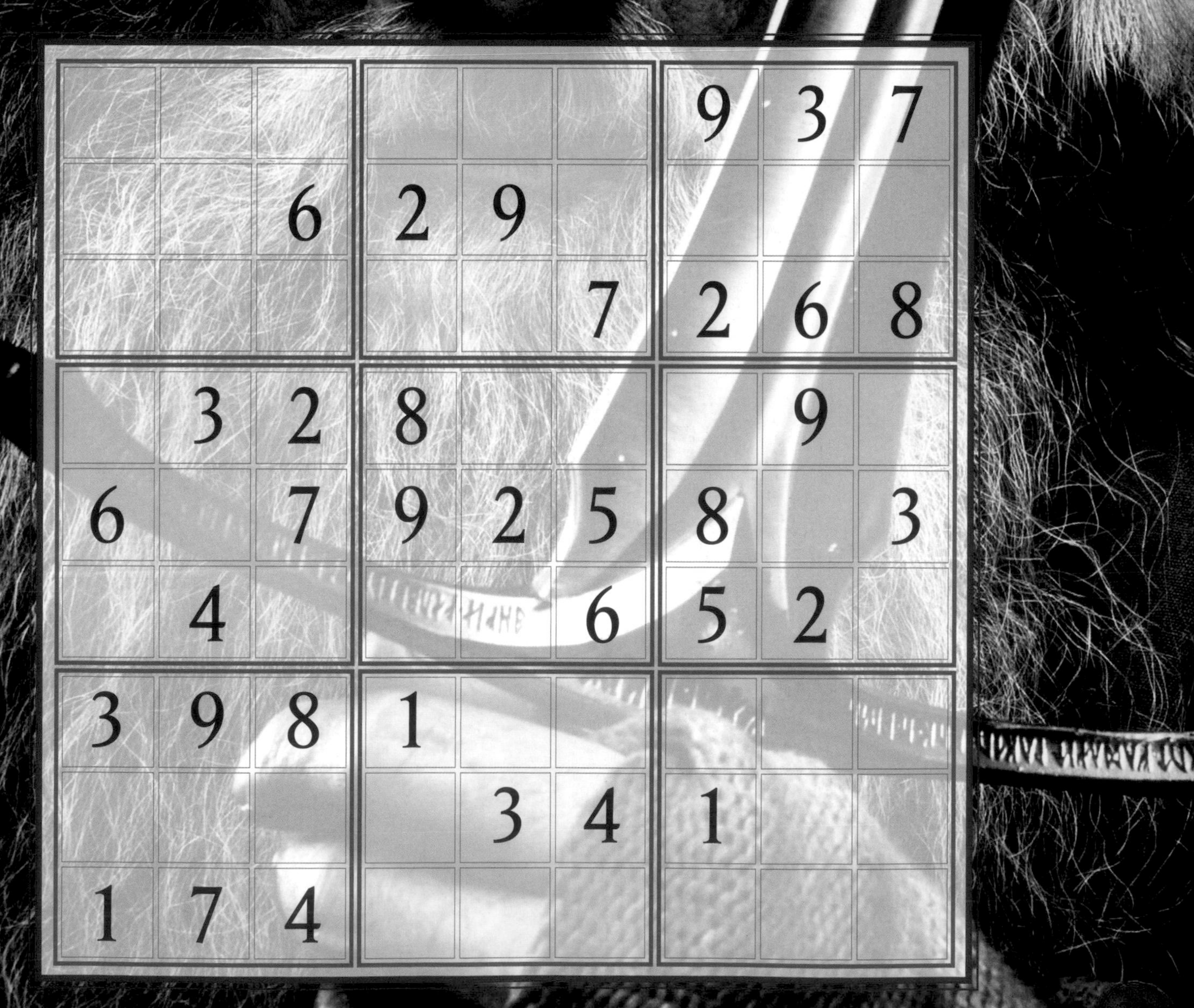

						9	3	7
		6	2	9				
					7	2	6	8
	3	2	8				9	
6		7	9	2	5	8		3
	4				6	5	2	
3	9	8	1					
				3	4	1		
1	7	4						

THE COMPANY OF DWARVES

DORI THE DWARF

Dori is the strongest of all the Dwarves and the elder brother of Nori and Ori. Dori never looks on the bright side of things and always thinks the worst is going to happen. He really cares about his brothers and spends most of his time looking out for Ori.

Dori and his brothers are distant relatives of Thorin. He might not always believe that the Company will succeed in their quest, but Dori would never give less than 100% effort to get the job done.

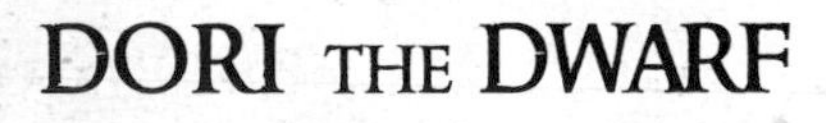

NORI THE DWARF

A Dwarf who spends most of his time in trouble, Nori is always up to something – and that something is usually illegal. An expert at picking locks, if something dubious needs doing, Nori is your Dwarf.

It is rumoured that the only reason Nori joined the quest in the first place is because he was on the run from trouble in his own town and needed to get away fast! He doesn't always get along with his brothers, but Nori will protect them until his dying breath.

DID YOU KNOW?

Elves and Dwarves really don't like each other that much!

ORI THE DWARF

The youngest of the three brothers, Ori is a brilliant artist. Like Bilbo, Ori keeps a journal and spends a lot of time writing and drawing in it. Most of the time Ori is relatively quiet and polite, but he has a surprising amount of courage and determination. He seems to spend a lot of his time being bossed about by his brothers, but it doesn't mean he won't stand up for himself when the time is right. Ori also happens to be quite a good shot with a catapult!

BIFUR THE DWARF

The first thing anyone notices about Bifur is usually the rusting remains of an Orc axe that is stuck in his head! Unable to talk, Bifur grunts and uses hand signals to communicate.

Unlike most of the others in the Company of Dwarves, Bifur is not directly related to Thorin. He is not of noble lineage, but is descended from miners and smithies.

CHOOSE YOUR WEAPON

The path to the Lonely Mountain is difficult, dangerous and deadly to those who are unarmed. Use these pictures to help design your own legendary weapon to wield in battle.

Dwarven weapons have sharp corners and angles on their blades and handles.

Like Orcrist, weapons designed by Elves are smoothly curved and elegant.

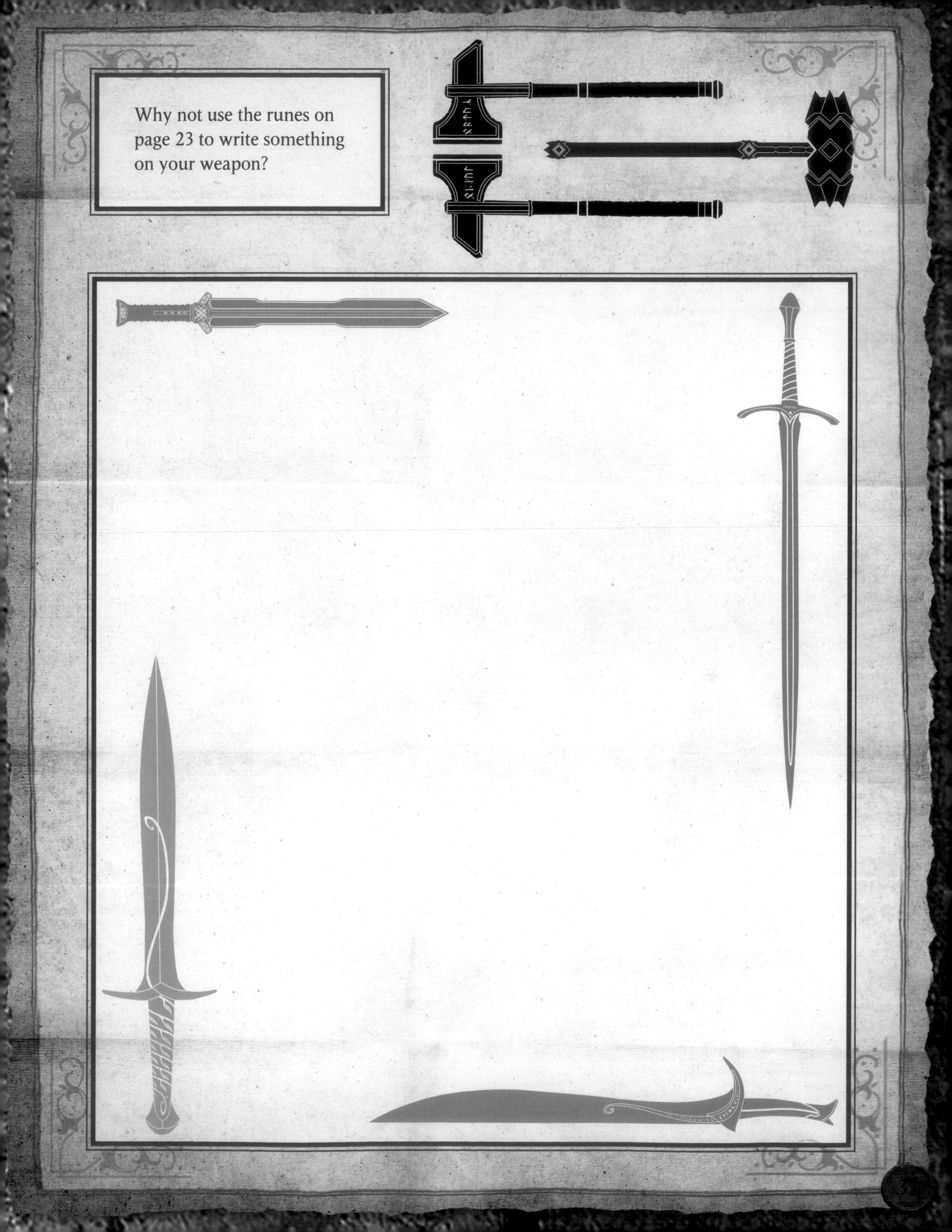
Why not use the runes on page 23 to write something on your weapon?

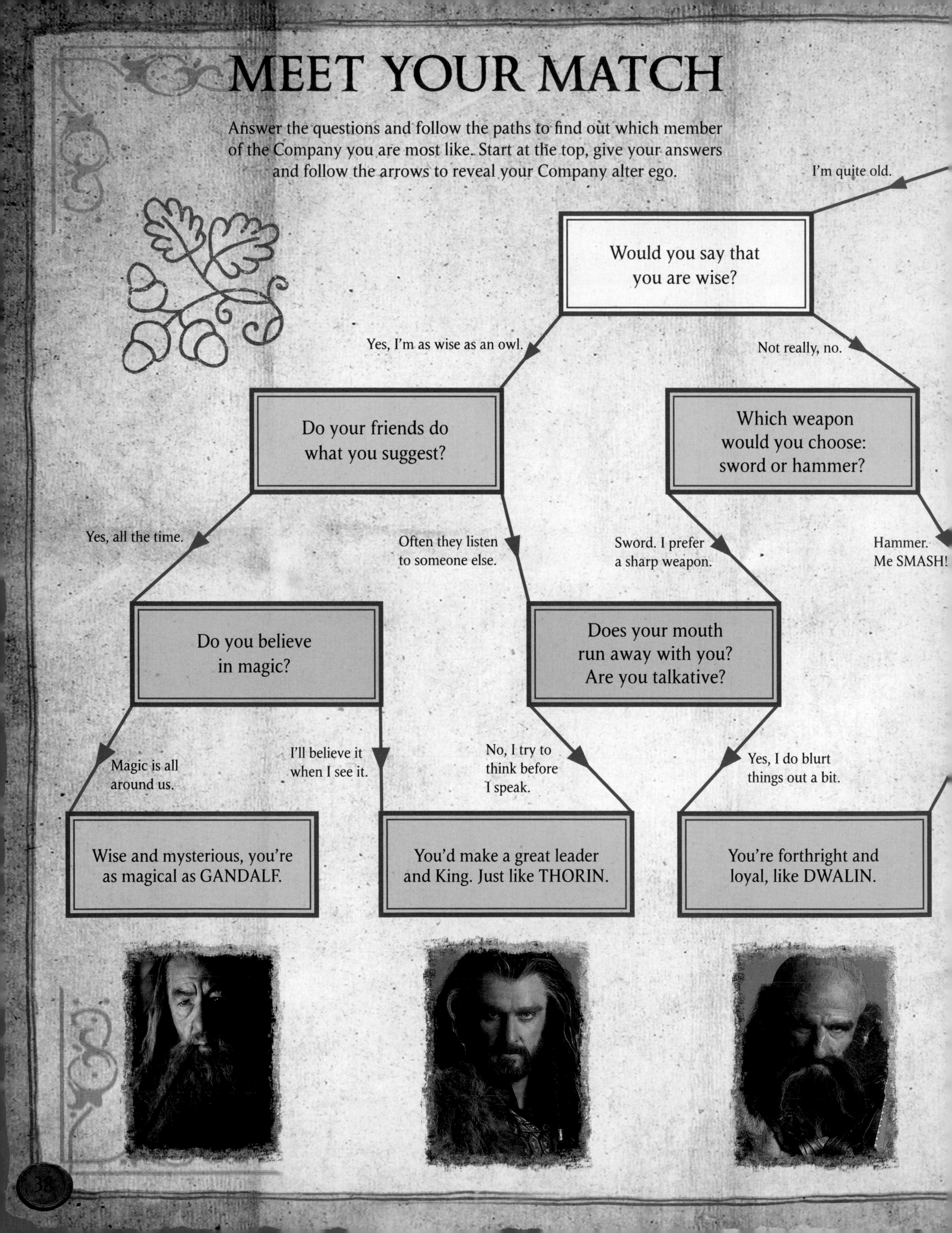

MEET YOUR MATCH

Answer the questions and follow the paths to find out which member of the Company you are most like. Start at the top, give your answers and follow the arrows to reveal your Company alter ego.

I'm quite old.

Would you say that you are wise?

Yes, I'm as wise as an owl.

Do your friends do what you suggest?

Yes, all the time.

Do you believe in magic?

Magic is all around us.

Wise and mysterious, you're as magical as GANDALF.

I'll believe it when I see it.

You'd make a great leader and King. Just like THORIN.

Often they listen to someone else.

Does your mouth run away with you? Are you talkative?

No, I try to think before I speak.

You'd make a great leader and King. Just like THORIN.

Yes, I do blurt things out a bit.

You're forthright and loyal, like DWALIN.

Not really, no.

Which weapon would you choose: sword or hammer?

Sword. I prefer a sharp weapon.

Hammer. Me SMASH!

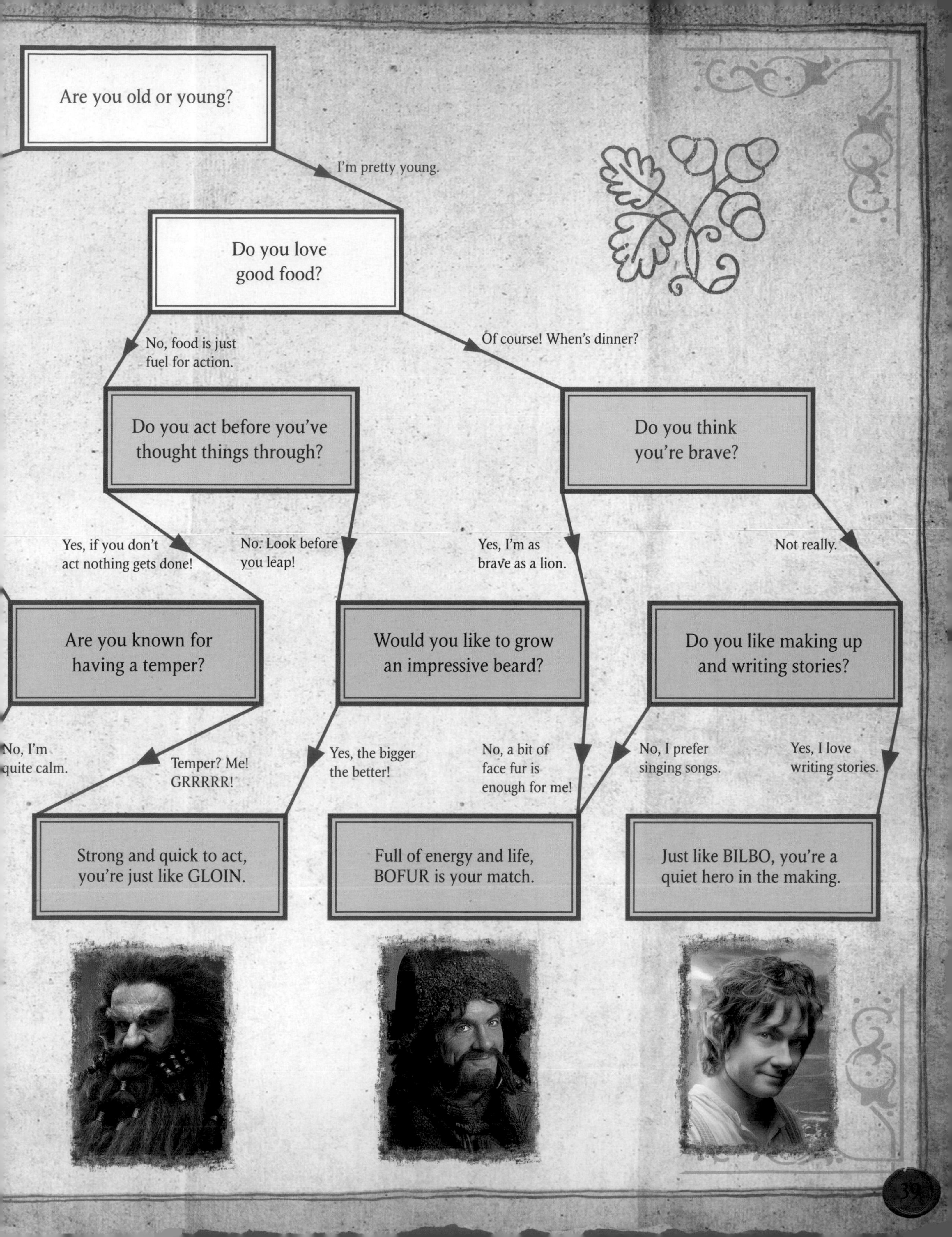

Are you old or young?
I'm pretty young.
Do you love good food?
No, food is just fuel for action.
Of course! When's dinner?
Do you act before you've thought things through?
Do you think you're brave?
Yes, if you don't act nothing gets done!
No. Look before you leap!
Yes, I'm as brave as a lion.
Not really.
Are you known for having a temper?
Would you like to grow an impressive beard?
Do you like making up and writing stories?
No, I'm quite calm.
Temper? Me! GRRRRR!
Yes, the bigger the better!
No, a bit of face fur is enough for me!
No, I prefer singing songs.
Yes, I love writing stories.
Strong and quick to act, you're just like GLOIN.
Full of energy and life, BOFUR is your match.
Just like BILBO, you're a quiet hero in the making.

TO ARMS!

Help the Dwarves get ready for battle by matching them to their weapons.

FILI

DWALIN

ORI

BOFUR

BOMBUR

A.

B.

D.

E.

F.

INTO BATTLE!

TO THE RESCUE

Goblins have captured Bilbo and the rest of the Company! Show Gandalf the way through the caves and past the guards to the finish so he can free his friends.

START

FINISH

DWARF DOUBLE

There are 10 differences between these pictures of Dori, Kili and Bifur. Can you spot them all? Circle them on the bottom picture.

A.

B.

Gollum

The Misty Mountains are home to many strange creatures – and none are more strange than Gollum. Pale and shrunken from years of living alone in the dark, Gollum seems quite mad when Bilbo first meets him.

Gollum is not this strange creature's real name. In fact Gollum is actually a hobbit-like creature called Sméagol! He has lived for so many years in the dark caves under the mountains that he doesn't look much like a hobbit any more though!

His 'precious' is an ordinary-looking gold ring and his prized possession. When Gollum thinks that Bilbo has stolen it, he flies into a rage.

DID YOU KNOW?

Gollum loves playing games, especially ones with riddles. In fact he loves games as much as he likes eating fish... or hobbitses.

Gollum lives alone in the middle of a vast underground lake. He knows the caves under the Misty Mountains better than anyone.

He has spent so long on his own that he has started to lose his mind with loneliness. In fact over the years he has become so lonely that he now talks to himself and even his 'precious' ring.

RIDDLES IN THE DARK

When Bilbo stumbles across Gollum in the dark caves under the Misty Mountains he must use his wits in a challenge of riddles. Are you clever enough to take the challenge too? Check your answers on page 61.

RIDDLE 1:

What has roots that nobody sees
Is taller than the trees
Up, up it goes,
And yet never grows?

SOLUTION:

RIDDLE 2:

Voiceless it cries,
Wingless flutters,
Toothless bites,
Mouthless mutters

SOLUTION:

RIDDLE 3:

It cannot be seen, cannot be felt,
Cannot be heard, cannot be smelt.
It lies behind stars and under hills,
And empty holes it fills,
It comes first and follows after,
Ends life, kills laughter.

SOLUTION:

RIDDLE 4:

A box without hinges,
key, or lid,
Yet golden treasure
inside is hid.

SOLUTION:

____

RIDDLE 5:

Alive without breath,
As cold as death;
Never thirsty, ever drinking,
All in mail, never clinking.

SOLUTION:

____

RIDDLE 6:

This thing all things devours:
Beasts, birds, trees, flowers;
Gnaws iron, bites steel;
Grinds hard stones to meal;
Slays kings, ruins town,
And beats high mountain down.

SOLUTION:

____

BILBO'S ESCAPE

Bilbo has stolen Gollum's precious and is trying to escape from the Misty Mountains. Get a counter and a die and see if you can help him find his way out of the caves before Gollum catches up.

1\. START

2\.

3\. Ooops. You stumble in the dark. Fill in 2 extra boxes while you get up.

4\.

5\.

6\.

7\. Sting is glowing! Fill in 1 more box while you stop to cover it up so it doesn't give you away.

8\.

9\.

10\.

11\.

RULES:

Roll the die and move your counter. Put a cross in a box here every time you roll the die. If you fill them all in before you reach freedom, then Gollum has caught you and you're lunch!

To play again, simply draw your own boxes on a piece of paper.

COUNTDOWN

15. Slip on the Ring and disappear! Fill in 1 box and jump to number 17.

16.

17.

19. Bother! Your buttons are stuck on the door. Fill in 1 space while you squeeze out.

20. FREEDOM! You've escaped Gollum. But there's still a long way to go until you reach the Lonely Mountain!

FINISH

14.

18. Take a giant leap over Gollum. Fill in 1 box and jump to number 20.

13.

12. Stop to hide from a Goblin patrol. Fill in 2 more boxes.

Gandalf

THE LOST KEY

Alas, the Goblin King has taken Thorin's key to Erebor from him. Can you select the correct key from the six shadows below?

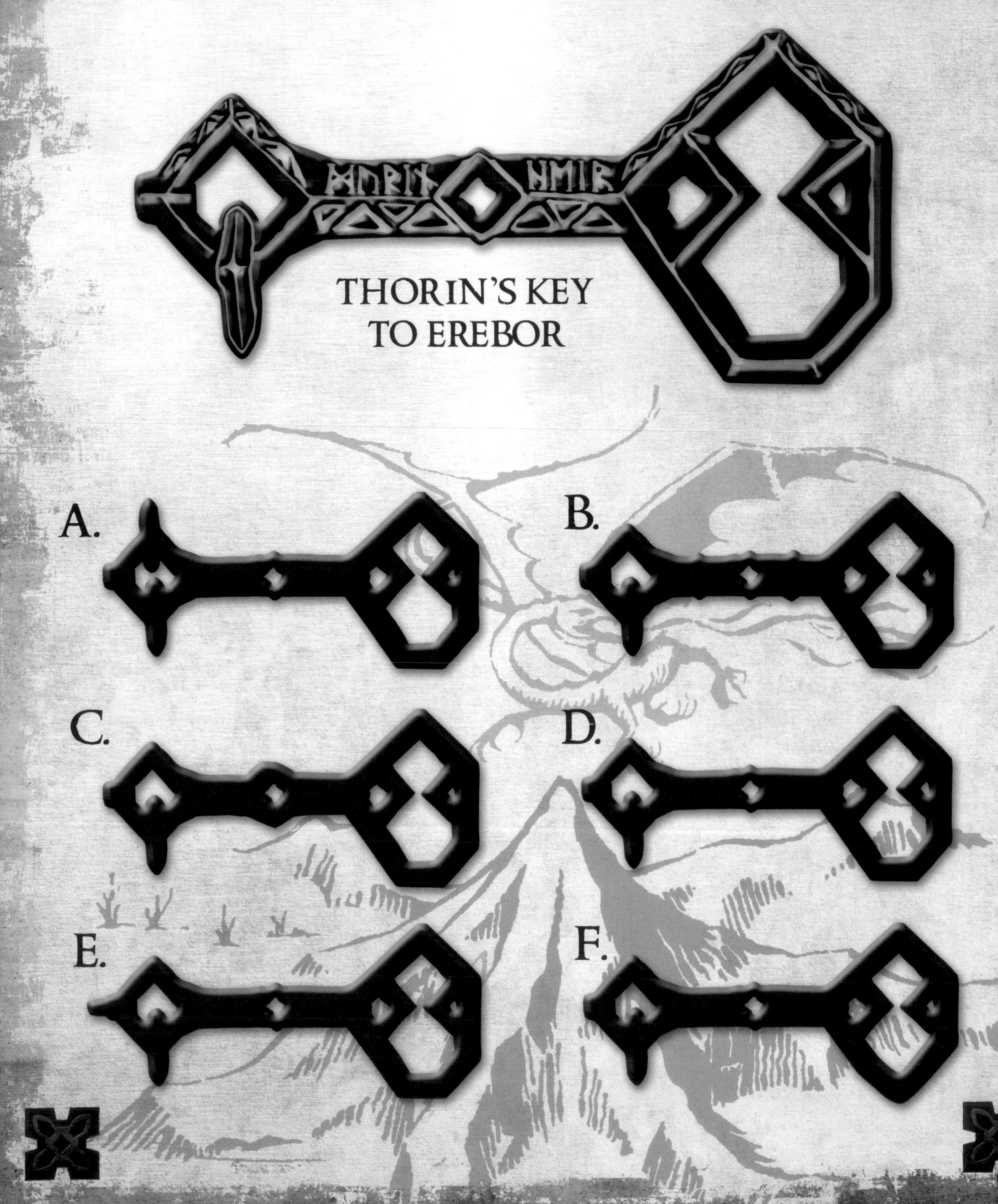

Thranduil

An old and powerful Elf, Thranduil is a cunning warrior who has fought in many battles. As the Elvenking he rules the Wood Elves who live in Mirkwood. He is also the father of an Elf called Legolas who will go on to have his own great adventures with Bilbo's cousin, Frodo.

Like most Elves, Thranduil doesn't really like Dwarves – especially Thorin and his Company. In fact, when he finds Thorin and his friends wandering in Mirkwood, he orders his Wood Elves to take them prisoner.

Thranduil knows how to throw a party and one of his celebrations gives Bilbo the perfect chance to help free his friends and escape from Mirkwood.

Unlike many creatures who pass through Mirkwood, Thranduil and his Wood Elves are not scared of the fearsome monsters that lurk in the shadows. Indeed, if it weren't for Thranduil's people saving them, Thorin and his companions would surely have died deep in the woods.

ON A ROLL

Bilbo has had a cunning plan to help the Company escape. Help him work out which path to roll the barrels along to find a way out of the Halls of the Elven King.

A. B. C.

ESCAPE THE WARGS!

A vicious pack of Wargs approach and poor Bilbo has no place to hide! Be quick and help him find his way to Bofur, who can help him to safety.

START

FINISH

ORI'S ART

Ori needs a picture of Bilbo for his journal.
Look at the picture on the left and copy it square by square into the larger grid below to complete the portrait.

ANSWERS

PANTRY PUZZLE page 10

BILBO'S BOTTLES page 16

SACK SEARCH page 20

R	B	H	E	E	F	O	M	R	H	N	I	N	L	E	D
L	U	S	N	I	C	I	R	R	D	W	I	M	M	O	W
I	O	B	L	S	W	S	E	C	H	E	N	O	C	H	A
N	I	I	M	O	R	M	T	N	T	U	U	R	A	I	L
G	I	E	R	O	K	B	O	R	W	I	N	I	T	S	I
A	L	L	V	C	B	L	D	R	D	N	H	E	R	I	N
R	C	O	I	M	I	O	O	M	L	O	E	R	T	E	L
H	S	W	I	W	D	E	R	Y	O	T	R	F	T	S	M
Y	G	E	T	N	G	I	D	W	E	O	H	I	E	T	S
L	A	N	I	R	O	H	T	S	A	L	O	H	E	T	F
T	C	F	E	T	M	E	B	A	L	I	N	M	T	S	O
R	U	F	I	B	P	I	I	O	T	T	M	Z	G	R	I
O	Y	O	E	K	P	N	N	I	L	I	K	A	W	A	F
M	S	T	O	E	O	T	E	H	E	O	C	O	R	T	N
I	O	E	C	R	E	S	O	B	Z	O	B	O	F	U	R
S	R	L	I	T	D	V	U	M	E	D	T	S	E	O	C

MOON RUNES page 23

Stand by the grey stone when the thrush knocks and the setting sun with the last light of durins day will shine upon the keyhole.

PICTURE PIECES pages 26-27

1=F, 2=B, 3=G, 4=H

A HELPING HAND pages 30-31

PUZZLE 1. A =1, B=3, C=2, D=4

PUZZLE 2. Fill the 5 litre jar completely. Pour the water from the 5 litre jar into the 3 litre jar until it is full. There are 2 litres left in the 5 litre jar. Pour away the water from the 3 litre jar. Then pour the 2 litres from the 5 litre jar into the 3 litre jar. Completely fill the 5 litre jar again. Pour water from the 5 litre jar into the 3 litre jar until it is full. This will only take 1 litre, which will leave 4 litres in the 5 litre jar.

PUZZLE 3.

			38	87	49			
		11	22	66	22	11		
	9	12	6	54	6	12	9	
7	3	5	2	34	2	5	3	7

PUZZLE 4. The answer is 35 (the bottom number is the product of multiplying the other 2 numbers together).

PUZZLE 5. The answer is 2 potions. If you take 2, you have 2!

MAGIC SQUARES page 33

4	2	1	5	6	8	9	3	7
7	8	6	2	9	3	4	1	5
9	5	3	4	1	7	2	6	8
5	3	2	8	4	1	7	9	6
6	1	7	9	2	5	8	4	3
8	4	9	3	7	6	5	2	1
3	9	8	1	5	2	6	7	4
2	6	5	7	3	4	1	8	9
1	7	4	6	8	9	3	5	2

TO ARMS! page 40

Thorin = F, Fili = A, Dwalin = B, Ori = C, Bofur = E, Bombur = D.

TO THE RESCUE pages 44-45

DWARF DOUBLE page 46

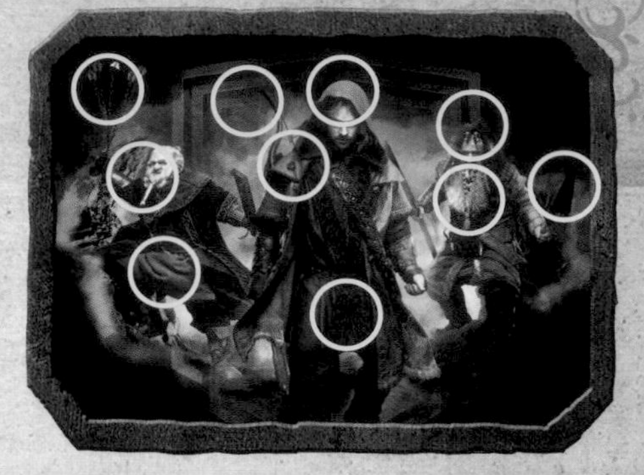

RIDDLES IN THE DARK pages 48-49

SOLUTION 1: A mountain

SOLUTION 2: The wind

SOLUTION 3: The dark

SOLUTION 4: An egg

SOLUTION 5: A fish

SOLUTION 6: Time

THE LOST KEY page 53

The correct key is D.

ON A ROLL page 55

The right path is C.

ESCAPE THE WARGS! page 56

AN UNEXPECTED QUIZ pages 58-59

1. Bag End, 2. Boing, 3. Thror's Key, 4. Glamdring, 5. Gimli, 6. Bifur, 7. Smaug, 8. The Goblin King, 9. Narsil, 10. A ring

Middle
-earth
Ice Bay
Carn Dûm
Angmar
Arnor
Eriador
Hills of
Evendim
Lake
Evendim
Fornost
Annúminas
Forlindon
Blue Mountains
Lune River
Bindbale Wood
Needlehole
Dwaling
Scary
Brockenborings
Rushock Bog
Overhill
The Hill
Little Delving
The Water
Michel
Delving
Hobbiton
Bywater
The Shire
Longbottom
Tower Hills
Forlond
Gulf of Lune
The Grey
Havens
Harlond
Harlindon
Girdley Island
Buckland
Chetwood
Bree
Barrow-downs
Old Forest
The Greenway
Shirebourn
Overbourn
Marshes
Sarn Ford
Brandywine River
Weather
Hills
Weathertop
The Great East Road
The Last
Bridge
South
Downs
Tharbad
Swanfleet River
Minhiriath
Greyflood
River
Eryn
Vorn
Enedwaith
Old South Road
Dunland
Isen
River
Adorn River
Drúwaith